Livestock Production and Management

Livestock Production and Management

Edited by
Bran Powell

Larsen & Keller
www.larsen-keller.com

Livestock Production and Management
Edited by Bran Powell
ISBN: 978-1-63549-166-1 (Hardback)

⊟ Larsen & Keller

Published by Larsen and Keller Education,
5 Penn Plaza,
19th Floor,
New York, NY 10001, USA

Cataloging-in-Publication Data

Livestock production and management / edited by Bran Powell.
    p. cm.
Includes bibliographical references and index.
ISBN 978-1-63549-166-1
1. Livestock. 2. Livestock--Breeding. 3. Animal culture. 4. Livestock--Handling. I. Powell, Bran.
SF77 .L58 2017
636--dc23

The publisher's policy is to use permanent paper from mills that operate a sustainable forestry policy. Furthermore, the publisher ensures that the text paper and cover boards used have met acceptable environmental accreditation standards.

Printed and bound in the United States of America.

For more information regarding Larsen and Keller Education and its products, please visit the publisher's website www.larsen-keller.com

Table of Contents

Preface

Livestock refers to the practice of raising domesticated animals in order to produce food, labour and fiber. Livestock are raised ad selectively bred for promoting favorable traits that contribute to the food industry. In this book, we will talk thoroughly about the various concepts of livestock production and management. It will also give detailed explanations about the practices of animal welfare and environmental impact. Some of the diverse topics covered in the book address the different branches that fall under this category. Those in search to broaden their knowledge about livestock management will be greatly assisted by this book.

A foreword of all chapters of the book is provided below:

Chapter 1 - Animals which are domesticated and used to produce commodities such as food and fiber are known as livestock. This chapter broadly introduces the reader to the term livestock. The section on livestock offers an insightful focus, keeping in mind the complex subject matter.; **Chapter 2** - The term livestock can be explained both broadly and narrowly. Livestock can broadly be divided into domestic and captive animals, and some of the animals listed in this chapter include goat, sheep, domestic duck and cattle. This chapter explicates on captivity, which helps in broadening the existing knowledge on livestock.; **Chapter 3** - The management of farm animals by humans is known as animal husbandry while intensive animal farming means to keep livestock at higher stocking densities than usual. The main purpose of maintaining livestock is to produce eggs, meat and milk for human consumption. This chapter explains the maintenance of livestock by giving a brief description on animal husbandry and intensive animal farming. ; **Chapter 4** - Humans, for a number of reasons and purposes use livestock. Livestock products include fiber, fur, honey, dairy products etc. Modern techniques seek to minimize human involvement while increasing outcome and improving animal health. The following content helps the reader to develop a better understanding on the commodities produced by livestock.; **Chapter 5** - Droving is the practice of moving your livestock by making them walk over long distances. Droving can be traced back to the ancient cultures, where it was necessary to source food from different cities. This chapter also elucidates on livestock transportation and its relation with auction, livestock shows, slaughter and selective breeding.; **Chapter 6** - The fields covered in this chapter are poultry, sericulture and beekeeping. Poultry is the keeping of birds for human purposes such as the production of eggs and meat whereas the maintenance of bees for honey is beekeeping. This section explains to the reader the relation between livestock and its allied fields. It is a compilation of the various branches of livestock production that form an integral part of the broader subject matter.

At the end, I would like to thank all the people associated with this book devoting their precious time and providing their valuable contributions to this book. I would also like to express my gratitude to my fellow colleagues who encouraged me throughout the process.

Editor

Introduction to Livestock

Animals which are domesticated and used to produce commodities such as food and fiber are known as livestock. This chapter broadly introduces the reader to the term livestock. The section on livestock offers an insightful focus, keeping in mind the complex subject matter.

Livestock are domesticated animals raised in an agricultural setting to produce commodities such as food, fiber, and labor. The term is often used to refer solely to those raised for food, and sometimes only farmed ruminants, such as cattle and goats. In recent years, some organizations have also raised livestock to promote the survival of rare breeds. The breeding, maintenance, and slaughter of these animals, known as animal husbandry, is a component of modern agriculture that has been practiced in many cultures since humanity's transition to farming from hunter-gatherer lifestyles.

Animal husbandry practices have varied widely across cultures and time periods. Originally, livestock were not confined by fences or enclosures, but these practices have largely shifted to intensive animal farming, sometimes referred to as "factory farming". These practices increase yield of the various commercial outputs, but have led to increased concerns about animal welfare and environmental impact. Livestock production continues to play a major economic and cultural role in numerous rural communities.

Etymology and Legal Definition

Livestock as a word was first used between 1650 and 1660, as a merger between the words "live" and "stock".

Older English sources, such as the King James Version of the Bible, refer to all domesticated animals as "cattle", while the word "deer" was used for wild animals. The word cattle is derived from Old North French *catel*, which meant all kinds of movable personal property, including livestock, which was differentiated from immovable real estate ("real property"). In later English, sometimes smaller livestock such as chickens and pigs were referred to as "small cattle". Today, the modern meaning of cattle, without a modifier, usually refers to domesticated bovines, but sometimes livestock refers only to this subgroup.

Legal definition

United States federal legislation sometimes more narrowly defines the term to make specified agricultural commodities either eligible or ineligible for a program or activity.

For example, the Livestock Mandatory Reporting Act of 1999 (P.L. 106-78, Title IX) defines livestock only as cattle, swine, and sheep. The 1988 disaster assistance legislation defined the term as "cattle, sheep, goats, swine, poultry (including egg-producing poultry), equine animals used for food or in the production of food, fish used for food, and other animals designated by the Secretary."

History

Animal-rearing originated during the cultural transition to settled farming communities from hunter-gatherer lifestyles. Animals are domesticated when their breeding and living conditions are controlled by humans. Over time, the collective behaviour, lifecycle and physiology of livestock have changed radically. Many modern farm animals are unsuited to life in the wild.

Dogs were domesticated in East Asia about 15,000 years ago. Goats and sheep were domesticated around 8000 BC in Asia. Swine or pigs were domesticated by 7000 BC in the Middle East and China. The earliest evidence of horse domestication dates to around 4000 BC.

Types

The term "livestock" is nebulous and may be defined narrowly or broadly. Broadly, livestock refers to any breed or population of animal kept by humans for a useful, commercial purpose. This can mean domestic animals, semidomestic animals, or captive wild animals. Semidomesticated refers to animals which are only lightly domesticated or of disputed status. These populations may also be in the process of domestication. Some people may use the term livestock to refer to only animals used for red meat.

Animal Rearing

A Brown Swiss cow in the Swiss Alps

Livestock are used by humans for a variety of purposes, many of which have an economic value. Livestock products include:

Meat

> A useful form of dietary protein and energy, meat is the edible tissue of the animal carcass.

Dairy products

> Mammalian livestock can be used as a source of milk, which can in turn easily be processed into other dairy products, such as yogurt, cheese, butter, ice cream, kefir, and kumis. Using livestock for this purpose can often yield several times the food energy of slaughtering the animal outright.

Clothing and adornment

> Livestock produce a range of fiber textiles. For example, domestic sheep and goats produce wool and mohair, respectively; cattle, swine, deer, and sheep skins can be made into leather; livestock bones, hooves and horns can be used to fabricate jewellery, pendants, or headgear.

Fertilizer

> Manure can be spread on fields to increase crop yields. This is an important reason why historically, plant and animal domestication have been intimately linked. Manure is also used to make plaster for walls and floors, and can be used as a fuel for fires. The blood and bone of animals are also used as fertilizer.

Labor

> The muscles of animals such as horses, donkeys, and yaks can be used to provide mechanical work. Prior to steam power, livestock were the only available source of nonhuman labor. They are still used in many places of the world to plough fields (drafting), transport goods and people, in military functions, and to power treadmills for grinding grain.

Land management

> The grazing of livestock is sometimes used as a way to control weeds and undergrowth. For example, in areas prone to wildfires, goats and sheep are set to graze on dry scrub which removes combustible material and reduces the risk of fires.

Conservation

> The raising of livestock to conserve a rare breed can be achieved through gene banking and breeding programmes.

During the history of animal husbandry, many secondary products have arisen in an attempt to increase carcass utilization and reduce waste. For example, animal offal and inedible parts may be transformed into products such as pet food and fertilizer. In the past, such waste products were sometimes also fed to livestock, as well. However, intraspecies recycling poses a disease risk, threatening animal and even human health. Due primarily to BSE (mad cow disease), feeding animal scraps to animals has been banned in many countries, at least for ruminants.

Farming Practices

Farming practices vary dramatically worldwide and among types of animals. Livestock are generally kept in an enclosure, fed by humans, and intentionally bred. However, some livestock are not enclosed, are fed by access to natural foods, and are allowed to breed freely.

Goat family with 1-week-old kid

Farrowing site in a natural cave in northern Spain

Historically, raising livestock was part of a nomadic or pastoral form of material culture. The herding of camels and reindeer in some parts of the world remains dissociated from sedentary agriculture. The transhumance form of herding in the California Sierra Nevada still continues, as cattle, sheep, or goats are moved from winter pasture in lower-elevation valleys to spring and summer pasture in the foothills and alpine regions, as the seasons progress. Cattle were raised on the open range in the western United States and Canada, on the Pampas of Argentina, and on other prairie and steppe regions of the world.

The enclosure of livestock in pastures and barns is a relatively new development in the history of agriculture. When cattle are enclosed, the type of confinement may vary from a small crate, a large-area fenced-in pasture, or a paddock. The type of feed may vary from naturally growing grass to animal feed. Animals are usually intentionally bred through artificial insemination or supervised mating. Indoor production systems are typically used for pigs, dairy cattle, poultry, veal cattle, dairy goats, and other animals depending on the region and season. Animals kept indoors are generally farmed intensively, as large space requirements could make indoor farming unprofitable if not impossible. However, indoor farming systems are controversial due to problems associated with handling animal waste, odours, the potential for groundwater contamination, and animal welfare concerns. (For a further discussion on intensively farmed livestock, see factory farming, and intensive pig farming). Livestock source verification is used to track livestock.

Other livestock are farmed outdoors, where the size of enclosures and the level of supervision may vary. In large, open ranges, animals may be only occasionally inspected or yarded in "round-ups" or a muster. Herding dogs may be used for mustering livestock, as are cowboys, stockmen, and jackaroos on horses, in vehicles, and in helicopters. Since the advent of barbed wire (in the 1870s) and electric fence technology, fencing pastures has become much more feasible and pasture management simplified. Rotation of pasturage is a modern technique for improving nutrition and health while avoiding environmental damage to the land. In some cases, very large numbers of animals may be kept in indoor or outdoor feeding operations (on feedlots), where the animals' feed is processed either offsite or onsite, and stored on site before being fed to the animals.

Livestock—-especially cattle—-may be branded to indicate ownership and age, but in modern farming, identification is more likely to be indicated by means of ear tags and electronic identification, instead. Sheep are also frequently marked by means of ear marks and/or ear tags. As fears of BSE and other epidemic illnesses mount, the use of implants to monitor and trace animals in the food production system is increasingly common, and sometimes required by government regulations.

Modern farming techniques seek to minimize human involvement, increase yield, and improve animal health. Economics, quality, and consumer safety all play roles in how animals are raised. The use of hard and soft drugs and feed supplements (or even feed type) may be regulated, or prohibited, to ensure that yield is not increased at the expense of consumer health, safety, or animal welfare. Practices vary around the world, for example growth hormone use is permitted in the United States, but not in stock to be sold in the European Union. The improvement of animal health using modern farming techniques has come into question. Feeding corn to cattle, which have historically eaten grasses, is an example; where the cattle are less adapted to this change, the rumen pH becomes more acidic, leading to liver damage and other health problems. The US Food and Drug Administration allows nonruminant animal proteins to be fed to cattle enclosed in feedlots. For example, it is acceptable to feed chicken manure and poultry meal to cattle, and beef or pork meat and bone meal to chickens.

Predation

Livestock farmers have suffered from wild animal predation and theft by rustlers. In North America, animals such as the gray wolf, grizzly bear, cougar, and coyote are sometimes considered a threat to livestock. In Eurasia and Africa, predators include the wolf, leopard, tiger, lion, dhole, Asiatic black bear, crocodile, spotted hyena, and other carnivores. In South America, feral dogs, jaguar, anacondas, and spectacled bears are threats to livestock. In Australia, the dingo, fox, and wedge-tailed eagle are common predators, with an additional threat from domestic dogs that may kill in response to a hunting instinct, leaving the carcass uneaten.

Disease

Livestock diseases compromise animal welfare, reduce productivity, and can infect humans. Animal diseases may be tolerated, reduced through animal husbandry, or reduced through antibiotics and vaccines. In developing countries, animal diseases are tolerated in animal husbandry, resulting in considerably reduced productivity, especially given the low health-status of many developing country herds. Disease management to improve productivity is often the first step taken in implementing an agriculture policy.

Disease management can be achieved by modifying animal husbandry practices. These measures aim to prevent infection with biosecurity measures such as controlling animal mixing and entry to farm lots, wearing protective clothing, and quarantining sick animals. Diseases also may be controlled by the use of vaccines and antibiotics. Antibiotics in subtherapeutic doses may also be used as a growth promoter, sometimes increasing growth by 10-15%. Concerns about antibiotic resistance have led in some cases to discouraging the practice of preventive dosing such as the use of antibiotic-laced feed. Countries often require veterinary certificates as a condition for transporting, selling, or exhibiting animals. Disease-free areas often rigorously enforce rules for preventing the entry of potentially diseased animals, including quarantine.

Transportation and Marketing

Since many livestock are herd animals, they were historically driven to market "on the hoof" to a town or other central location. During the period after the American Civil War, the abundance of Longhorn cattle in Texas and the demand for beef in Northern markets led to the implementation of the Old West cattle drive. The method is still used in some parts of the world. Truck transport is now common in developed countries. Local and regional livestock auctions and commodity markets facilitate trade in livestock. In other areas, livestock may be bought and sold in a bazaar, such as may be found in many parts of Central Asia, or in an informal flea market-type setting.

Grass-fed cattle, saleyards, Walcha, New South Wales

In developing countries, providing access to markets has encouraged farmers to invest in livestock, with the result being improved livelihoods. The International Crops Research Institute for the Semi-Arid Tropics (ICRISAT) has worked in Zimbabwe to help farmers make their most of their livestock herds. ICRISAT works to improve local farming systems through 'innovation platforms' at which farmers, traders, rural development agencies, and extension officers can discuss the challenges they faced. One finding was that if farmers devoted half of three hectares to maize and half to *mucuna* (velvet bean) in a rotation system, they could obtain 80% of the biomass needed to see their livestock through the dry season. If they only grew maize, they could only meet 20% of their biomass needs. In the town of Gwanda, the platform helped create a strong local market for goats, raising the value of a single animal from US$10 to $60. This gave the farmers a great incentive to invest in their own goats by growing their own feed stock, buying commercial feed only as a supplement, and improving their rangeland management techniques. Because the platform has helped regulate prices, farmers now plan ahead and sell animals at auction, rather than just selling one or two animals at their farm gate as opportunities arise.

Stock shows and fairs are events where people bring their best livestock to compete with one another. Organizations such as 4-H, Block & Bridle, and FFA encourage young people to raise livestock for show purposes. Special feeds are purchased and prior to the show, hours may be spent grooming the animal to look its best. In cattle, sheep, and swine shows, the winning animals are frequently auctioned off to the highest bidder, and the funds are placed into a scholarship fund for its owner. The movie *Grand Champion*, released in 2004, tells the story of a young Texas boy's experience raising a prize steer.

Animal Welfare

The issue of raising livestock for human benefit raises the issue of the suitable relationship between humans and animals, in terms of the status of animals and the obligations of people. The concept of animal welfare reflects the viewpoint that animals under human care should be treated in such a way that they do not suffer unnecessarily. What is considered 'unnecessary' suffering may vary. Generally, though, the animal welfare perspective is based on an interpretation of scientific research on farming practices. By contrast, animal rights defends the viewpoint that using animals for human benefit is, on principle, exploitation, regardless of the farming practices used. Animal rights activists are often vegan or vegetarian, whereas it is consistent with the animal welfare perspective to eat meat as long as the production processes are defensible.

A shepherd boy in India: Livestock are extremely important to the livelihoods of rural smallholder farmers, particularly in the developing world.

Animal welfare groups generally seek to generate public discussion on livestock raising practices and to secure greater regulation and scrutiny of livestock industry practices. Animal rights groups usually seek to abolish livestock farming, although some groups may recognise the necessity of first achieving more stringent regulation . Animal welfare groups such as the RSPCA are often, in first-world countries, given a voice at governmental level in the development of policy. Animal rights groups find it harder to convey their concerns, and as a result, may advocate civil disobedience or violence.

A number of animal husbandry practices have been the subject of campaigns in the 1990s and 2000s and have led to legislation in some countries. Confinement of livestock in small and unnatural spaces is often done for economic or health reasons. Animals may be kept in the minimum size of cage or pen with little or no space to exercise. Where livestock are used as a source of power, they may be pushed beyond their limits to the point of exhaustion. Increased public awareness and visibility of such abuse meant it was one of the first areas to receive legislation in the 19th century in European countries, but it continues in parts of Asia. Broiler hens may be debeaked, pigs may have deciduous teeth pulled, cattle may be dehorned and branded, dairy cows and sheep may have tails cropped, Merino sheep may undergo mulesing, and many types of male animals may be castrated. Animals may be transported long distances to market and slaughter, often under overcrowded conditions, heat stress, lack of feed and water, and without rest breaks. Such practices have been subject to legislation and protest. Appropriate methods to slaughter livestock were an early target for legislation. Campaigns continue to target halal and kosher religious ritual slaughter.

Environmental Impact

Cattle near the Bruneau River in Elko County, Nevada

Reports such as the United Nations report "Livestock's Long Shadow" cast a pall over the livestock sector (primarily cattle, chickens, and pigs) for 'emerging as one of the top two or three most significant contributors to our most serious environmental problems.' In April 2008, the United States Environmental Protection Agency released a major stock-taking of emissions in the United States entitled *Inventory of U.S. Greenhouse Gas Emissions and Sinks: 1990-2006*. It found that "In 2006, the agricultural sector was responsible for emissions of 454.1 teragrams of CO_2 equivalent (Tg CO_2 Eq.), or 6 percent of total U.S. greenhouse gas emissions." By way of comparison, transportation in the US produces more than 25% of all emissions. In 2009, Worldwatch Institute released a report which revealed that 51% of greenhouse gas emissions came from the animal agriculture sector.

The issue of livestock as a major policy focus remains, especially when dealing with problems of deforestation in neotropical areas, land degradation, climate change and air pollution, water shortage and water pollution, and loss of biodiversity. A research team at Obihiro University of Agriculture and Veterinary Medicine in Hokkaidō found that supplementing animals' diets with cysteine, a type of amino acid, and nitrate can reduce the amount of methane gas produced without jeopardising the cattle's productivity or the quality of their meat and milk.

Deforestation

Deforestation impacts the carbon cycle, as well as the global and regional climate, and causes the habitat loss of many species. Forests that are sinks for the carbon cycle are lost through deforestation. Forests are either logged or burned to make room for mining activities or for grasslands, and often the area needed for such purposes is extensive. Deforestation can also create fragmentation, allowing the survival of only patches of habitat in which species can live. If these patches are distant and small, gene flow is reduced, habitat is altered, edge effects occur, and more opportunities for invasive species to intrude occur.

Land Degradation

Research from the University of Botswana in 2008 found that farmers' common practice of overstocking cattle to make up for drought losses made ecosystems more vulnerable and risked long-term damage to cattle herds by depleting scarce biomass. The study of the Kgatleng district of Botswana predicted that by 2050, the cycle of mild drought is likely to become shorter for the region (18 months instead of two years) due to climate change.

Climate Change and Air Pollution

Methane is one of the gasses emitted from livestock manure; it persists in the atmosphere for long periods of time and is a potent greenhouse gas, the second-most abundant after carbon dioxide. Though less methane than carbon dioxide is produced, its ability to warm the atmosphere is 25 times greater. Nitrous oxide, another gaseous byproduct of animal agriculture, is about 300 times more potent at trapping heat in the atmosphere. Animal agriculture contributes 65% of anthropogenic nitrous oxide emissions.

Testing Australian sheep for exhaled methane production (2001), CSIRO

Water Shortage

Livestock require water not only for their own consumption, but also for watering the crops needed to produce their feed. Grains are often used to feed livestock; about 50% of US grains and 40% of world grains are used for this purpose Grain and crop production in general require various amounts of water depending on the end product. For example, 100,000 liters of water are needed to yield a kilogram of grain-fed beef, compared to 900 liters for a kilogram of wheat.

Water Pollution

Fertilizers that often contain manure are used to grow crops (such as cereals and fodder) that contain phosphorus and nitrogen, 95% of which is estimated to be lost to the environment. Water pollution from agricultural runoff causes dead zones for plants and aquatic animals due to the lack of oxygen in the water. This lack of oxygen, known as eutrophication, is caused when organisms present in the water grow excessively and then later decompose, in the process using up the oxygen in the water. A prominent example is the Gulf of Mexico, where much of the nutrients in fertilizer used in the US Midwest is funneled down the Mississippi River into the Gulf, causing massive dead zones. Other pollutants not commonly considered are antibiotics and hormones. In southern Asia, vultures that consumed carcasses of livestock declined 95% due to their ingestion of the antibiotic known as diclofenac.

Alternatives

Researchers in Australia are looking into the possibility of reducing methane from cattle and sheep by introducing digestive bacteria from kangaroo intestines into livestock. Furthermore, as a means to conserve traditional livestock, cryoconservation of animal genetic resources have been put into action. (Cryoconservation is a practice that involves collecting genetic material and storing it in low temperatures with an intent of conserving a particular breed.)

In semiarid rangelands such as the Great Plains in the U.S., research has provided evidence that livestock can be beneficial to maintaining grassland habitats for big game species.

Economic and Social Benefits

The value of global livestock production in 2013 has been estimated at about 883 billion dollars, (constant 2005-2006 dollars). However, economic implications of livestock production extend further: to downstream industry (saleyards, abattoirs, butchers, milk processors, refrigerated transport, wholesalers, retailers, food services, tanneries, etc.), upstream industry (feed producers, feed transport, farm and ranch supply companies, equipment manufacturers, seed companies, vaccine manufacturers, etc.) and associated services (veterinarians, nutrition consultants, shearers, etc.).

Livestock provide a variety of food and nonfood products; the latter include leather, wool, pharmaceuticals, bone products, industrial protein, and fats. For many abattoirs, very little animal biomass may be wasted at slaughter. Even intestinal contents removed at slaughter may be recovered for use as fertilizer. Livestock manure helps maintain the fertility of grazing lands. Manure is commonly collected from barns and feeding areas to fertilize cropland. In some places, animal manure is used as fuel, either directly (as in some developing countries), or indirectly (as a source of methane for heating or for generating electricity). In regions where machine power is limited, some classes of livestock are used as draft stock, not only for tillage and other on-farm use, but also for transport of people and goods. In 1997, livestock provided energy for between an estimated 25 and 64% of cultivation energy in the world's irrigated systems, and that 300 million draft animals were used globally in small-scale agriculture.

Although livestock production serves as a source of income, it can provide additional economic values for rural families, often serving as a major contributor to food security and economic security. Livestock can serve as insurance against risk and is an economic buffer (of income and/or food supply) in some regions and some economies (e.g., during some African droughts). However, its use as a buffer may sometimes be limited where alternatives are present, which may reflect strategic maintenance of insurance in addition to a desire to retain productive assets. Even for some livestock owners in developed nations, livestock can serve as a kind of insurance. Some crop growers may produce livestock as a strategy for diversification of their income sources, to reduce risks related to weather, markets and other factors.

Many studies have found evidence of the social, as well as economic, importance of livestock in developing countries and in regions of rural poverty, and such evidence is not confined to pastoral and nomadic societies.

Social values in developed countries can also be considerable. For example, in a study of livestock ranching permitted on national forest land in New Mexico, USA, it was concluded that "ranching maintains traditional values and connects families to ancestral lands and cultural heritage", and that a "sense of place, attachment to land, and the value of preserving open space were common themes". "The importance of land and animals as means of maintaining culture and way of life figured repeatedly in permittee responses, as did the subjects of responsibility and respect for land, animals, family, and community."

In the US, profit tends to rank low among motivations for involvement in livestock ranching. Instead, family, tradition and a desired way of life tend to be major motivators for ranch purchase, and ranchers "historically have been willing to accept low returns from livestock production."

References

- "Agriculture: A Glossary of Terms, Programs, and Laws, 2005 Edition" (PDF). Archived from the original (PDF) on 2011-02-12. Retrieved 2011-12-10.

- "Oldest Known Pet Cat? 9,500-Year-Old Burial Found on Cyprus". News.nationalgeographic. com. 2010-10-28. Retrieved 2011-12-10.

- "2011 U.S. Greenhouse Gas Inventory Report | Climate Change - Greenhouse Gas Emissions | U.S. EPA". Epa.gov. 2006-06-28. Retrieved 2011-12-10.

- "Global warming breakthrough: way to stop cow gas - Unusual Tales - Specials". Smh.com.au. 2008-01-22. Retrieved 2011-12-10.

- Marsha Walton (2004-04-09). "CNN.com - Ancient burial looks like human and pet cat - Apr 9, 2004". Edition.cnn.com. Retrieved 2011-12-10.

- Simmons, Michael (2009-09-10). "Dogs seized for killing sheep - Local News - News - General - The Times". Victorharbortimes.com.au. Retrieved 2011-12-10.

- "feed (agriculture) :: Antibiotics and other growth stimulants - Britannica Online Encyclopedia". Britannica.com. Retrieved 2011-12-10.

2

Types of Livestock

The term livestock can be explained both broadly and narrowly. Livestock can broadly be divided into domestic and captive animals, and some of the animals listed in this chapter include goat, sheep, domestic duck and cattle. This chapter explicates on captivity, which helps in broadening the existing knowledge on livestock.

Captivity (Animal)

Animals that live under human care are in captivity. Captivity can be used as a generalizing term to describe the keeping of either domesticated animals (livestock and pets) or wild animals. This may include, for example, farms, private homes, zoos and laboratories. Keeping animals in human captivity and under human care can thus be distinguished between three primary categories according to the particular motives, objectives and conditions.

Animal husbandry

History

The domestication of animals is the oldest documented instance of keeping animals in captivity. This process eventually resulted in habituation of wild animal species to survive in the company of, or by the labor of, human beings. Domesticated species are those whose behaviour, life cycle, or physiology has been altered as a result of their breeding and living conditions under human control for multiple generations. Probably the earliest known domestic animal was the dog, likely as early as 15000 BC among hunter-gatherers in several locations.

Macaque in cage

Throughout history not only domestic animals as pets and livestock were kept in captivity and under human care, but also wild animals. Some were failed domestication attempts. Also, in past times, primarily the wealthy, aristocrats and kings collected wild animals for various reasons. Contrary to domestication, the ferociousness and natural behaviour of the wild animals were preserved and exhibited. Today's zoos claim other reasons for keeping animals under human care: conservation, education and science.

A critically endangered Mexican gray wolf is kept in captivity for breeding purposes.

Behavior of Animals in Captivity

Captive animals, especially those not domesticated, sometimes develop abnormal behaviours.

One type of abnormal behaviour is *stereotypical behaviors*, i.e. repetitive and apparently purposeless motor behaviors. Examples of stereotypical behaviours include pacing, self-injury, route tracing and excessive self-grooming. These behaviors are associated with stress and lack of stimulation. Many who keep animals in captivity, attempt to prevent or decrease stereotypical behavior by introducing novel stimuli, known as environmental enrichment.

A type of abnormal behavior shown in captive animals is self-injurious behavior (SIB). Self-injurious behavior indicates any activity that involves biting, scratching, hitting, hair plucking, or eye poke that may result in injuring oneself. Although its reported incidence is low, self-injurious behavior is observed across a range of primate species, especially when they experience social isolation in infancy. Self-bite involves biting one's own body—typically the arms, legs, shoulders, or genitals. Threat bite involves biting one's own body—typically the hand, wrist, or forearm—while staring at the observer, conspecific, or mirror in a threatening manner. Self-hit involves striking oneself on any part of the body. Eye poking is a behavior (widely observed in primates) that presses the knuckle or finger into the orbital space above the eye socket. Hair plucking is a jerking motion applied to one's own hair with hands or teeth, resulting in its excessive removal.

The proximal causes of self-injurious behavior have been widely studied in captive primates; either social or nonsocial factors can trigger this type of behavior. Social factors include changes in group composition, stress, separation from the group, approaches by or aggression from members of other groups, conspecific male individuals nearby, separation from females, and removal from the group. Social isolation, particularly disruptions of early mother-rearing experiences, is an important risk factor. Studies have suggested that, although mother-reared rhesus macaques still exhibit some self-injurious behaviors, nursery-reared rhesus macaques are much more likely to self-abuse than mother-reared ones. Nonsocial factors include the presence of a small cut, a wound or irritant, cold weather, human contact, and frequent zoo visitors. For example, a study has shown that zoo visitor density positively correlates with the number of gorillas banging on the barrier, and that low zoo visitor density caused gorillas to behave in a more relaxed way. Captive animals often cannot escape the attention and disruption caused by the general public, and the stress resulting from this lack of environmental control may lead to an increased rate of self-injurious behaviors.

On top of self inflicted harm, some animals exhibit harm towards others and internal psychological harm. This can be exhibited in various forma, such as Orca whales, which never have killed a human in the wild, killing two of its own trainers. Psychological tics can also be identified, ranging from swaying to head bobbing to pacing. Continuous inbreeding is also bringing out mental disadvantages, such as crossed eyes and infertility.

Studies suggest that many abnormal captive behaviors, including self-injurious behavior, can be successfully treated by pair housing. Pair housing provides a previously single-housed animal with a same-sex social partner; this method is especially effective with primates, which are widely known to be social animals. Social companionship provided by pair housing encourages social interaction, thus reducing abnormal and anxiety-related behavior in captive animals as well as increasing their locomotion.

Various Saptivity (Animal)

Goat

The domestic goat (*Capra aegagrus hircus*) is a subspecies of goat domesticated from the wild goat of southwest Asia and Eastern Europe.

The goat is a member of the family Bovidae and is closely related to the sheep as both are in the goat-antelope subfamily Caprinae. There are over 300 distinct breeds of goat. Goats are one of the oldest domesticated species, and have been used for their milk, meat, hair, and skins over much of the world. In 2011, there were more than 924 million live goats around the globe, according to the UN Food and Agriculture Organization.

Female goats are referred to as "does" or "nannies;" intact males are called "bucks" or "billies;" and juveniles of both sexes are called "kids". Castrated males are called "wethers". Goat meat from younger animals is called "kid" or *cabrito* (Spanish), while meat from older animals is known simply as "goat" or sometimes called *chevon*, or in some areas "mutton" (which more often refers to adult sheep meat).

Etymology

The Modern English word *goat* comes from Old English *gāt* "she-goat, goat in general", which in turn derives from Proto-Germanic **gaitaz* (cf. Dutch/Icelandic *geit*, German *Geiß*, and Gothic *gaits*), ultimately from Proto-Indo-European **□□aidos* meaning "young goat" (cf. Latin *haedus* "kid"), itself perhaps from a root meaning "jump" (assuming that Old Church Slavonic *zaję̌ci* "hare", Sanskrit *jihīte* "he moves" are related). To refer to the male, Old English used *bucca* (giving modern *buck*) until ousted by *hegote, hegoote* in the late 12th century. *Nanny goat* (females) originated in the 18th century and *billy goat* (for males) in the 19th.

History

Goats are among the earliest animals domesticated by humans. The most recent genetic analysis confirms the archaeological evidence that the wild Bezoar ibex of the Zagros Mountains is the likely original ancestor of probably all domestic goats today.

Horn cores from the Neolithic village of Atlit Yam

Neolithic farmers began to herd wild goats primarily for easy access to milk and meat, as well as to their dung, which was used as fuel, and their bones, hair and sinew for clothing, building and tools. The earliest remnants of domesticated goats dating 10,000 years before present are found in Ganj Dareh in Iran. Goat remains have been found at archaeological sites in Jericho, Choga Mami Djeitun and Çayönü, dating the domestication of goats in Western Asia at between 8000 and 9000 years ago.

Studies of DNA evidence suggests 10,000 years BP as the domestication date.

Historically, goat hide has been used for water and wine bottles in both traveling and transporting wine for sale. It has also been used to produce parchment.

Anatomy and Health

Goats are considered small livestock animals, compared to bigger animals such as cattle, camels and horses, but larger than microlivestock such as poultry, rabbits, cavies, and bees. Each recognized breed of goats has specific weight ranges, which vary from over 140 kg (300 lb) for bucks of larger breeds such as the Boer, to 20 to 27 kg (45 to 60 lb) for smaller goat does. Within each breed, different strains or bloodlines may have different recognized sizes. At the bottom of the size range are miniature breeds such as the African Pygmy, which stand 41 to 58 cm (16 to 23 in) at the shoulder as adults.

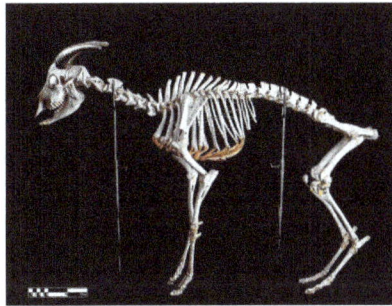

Skeleton (Capra hircus)

A white Irish goat with horns

Most goats naturally have two horns, of various shapes and sizes depending on the breed. Goats have horns unless they are "polled" (meaning, genetically hornless) or the horns have been removed, typically soon after birth. There have been incidents of polycerate goats (having as many as eight horns), although this is a genetic rarity thought to be inherited. The horns are most typically removed in commercial dairy goat herds, to reduce the injuries to humans and other goats. Unlike cattle, goats have not been successfully bred to be reliably polled, as the genes determining sex and those determining horns are closely linked. Breeding together two genetically polled goats results in a high number of intersex individuals among the offspring, which are typically sterile. Their horns are made of living bone surrounded by keratin and other proteins, and are used for defense, dominance, and territoriality.

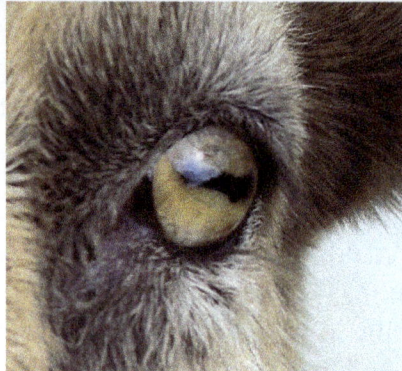

Eye with horizontal pupil

Goats are ruminants. They have a four-chambered stomach consisting of the rumen, the reticulum, the omasum, and the abomasum. As with other mammal ruminants, they are even-toed ungulates. The females have an udder consisting of two teats, in contrast to cattle, which have four teats. An exception to this is the Boer goat, which sometimes may have up to eight teats.

Goats have horizontal, slit-shaped pupils. Because goats' irises are usually pale, their contrasting pupils are much more noticeable than in animals such as cattle, deer, most horses and many sheep, whose similarly horizontal pupils blend into a dark iris and sclera.

Both male and female goats have beards, and many types of goat (most commonly dairy goats, dairy-cross Boers, and pygmy goats) may have wattles, one dangling from each side of the neck.

Some breeds of sheep and goats look similar, but they can usually be told apart because goat tails are short and usually point up, whereas sheep tails hang down and are usually longer and bigger – though some (like those of Northern European short-tailed sheep) are short, and longer ones are often docked.

Reproduction

Goats reach puberty between three and 15 months of age, depending on breed and nutritional status. Many breeders prefer to postpone breeding until the doe has reached 70% of the adult weight. However, this separation is rarely possible in extensively managed, open-range herds.

A two-month-old goat kid in a field of capeweed

In temperate climates and among the Swiss breeds, the breeding season commences as the day length shortens, and ends in early spring or before. In equatorial regions, goats are able to breed at any time of the year. Successful breeding in these regions depends more on available forage than on day length. Does of any breed or region come into estrus (heat) every 21 days for two to 48 hours. A doe in heat typically flags (vigorously wags) her tail often, stays near the buck if one is present, becomes more vocal, and may also show a decrease in appetite and milk production for the duration of the heat.

A female goat and two kids

Bucks (intact males) of Swiss and northern breeds come into rut in the fall as with the does' heat cycles. Bucks of equatorial breeds may show seasonal reduced fertility, but as with the does, are capable of breeding at all times. Rut is characterized by a decrease in appetite and obsessive interest in the does. A buck in rut will display flehmen lip curling and will urinate on his forelegs and face. Sebaceous scent glands at the base of the horns add to the male goat's odor, which is important to make him attractive to the female. Some does will not mate with a buck which has been descented.

In addition to natural, traditional mating, artificial insemination has gained popularity among goat breeders, as it allows easy access to a wide variety of bloodlines.

Gestation length is approximately 150 days. Twins are the usual result, with single and triplet births also common. Less frequent are litters of quadruplet, quintuplet, and even sextuplet kids. Birthing, known as kidding, generally occurs uneventfully. Just before kidding, the doe will have a sunken area around the tail and hip, as well as heavy breathing. She may have a worried look, become restless and display great affection for her keeper. The mother often eats the placenta, which gives her much-needed nutrients, helps stanch her bleeding, and parallels the behavior of wild herbivores, such as deer, to reduce the lure of the birth scent for predators.

Freshening (coming into milk production) occurs at kidding. Milk production varies with the breed, age, quality, and diet of the doe; dairy goats generally produce between 680 and 1,810 kg (1,500 and 4,000 lb) of milk per 305-day lactation. On average, a good quality dairy doe will give at least 3 kg (6 lb) of milk per day while she is in milk. A first-time milker may produce less, or as much as 7 kg (16 lb), or more of milk in exceptional cases. After the lactation, the doe will "dry off", typically after she has been bred. Occasionally, goats that have not been bred and are continuously milked will continue lactation beyond the typical 305 days. Meat, fiber, and pet breeds are not usually milked and simply produce enough for the kids until weaning.

Male lactation is also known to occur in goats.

Diet

Goats are reputed to be willing to eat almost anything, including tin cans and cardboard boxes. While goats will not actually eat inedible material, they are browsing animals, not grazers like cattle and sheep, and (coupled with their highly curious nature) will chew on and taste just about anything remotely resembling plant matter to decide whether it is good to eat, including cardboard, clothing and paper (such as labels from tin cans). The unusual smells of leftover food in discarded cans or boxes may further stimulate their curiosity.

A domestic goat feeding in a field of capeweed, a weed which is toxic to most stock animals

Aside from sampling many things, goats are quite particular in what they actually consume, preferring to browse on the tips of woody shrubs and trees, as well as the occasional broad-leaved plant. However, it can fairly be said that their plant diet is extremely varied, and includes some species which are otherwise toxic. They will seldom consume soiled food or contaminated water unless facing starvation. This is one reason goat-rearing is most often free ranging, since stall-fed goat-rearing involves extensive upkeep and is seldom commercially viable.

Goats prefer to browse on vines, such as kudzu, on shrubbery and on weeds, more like deer than sheep, preferring them to grasses. Nightshade is poisonous; wilted fruit tree leaves can also kill goats. Silage (fermented corn stalks) and haylage (fermented grass hay) can be used if consumed immediately after opening – goats are particularly sensitive to *Listeria* bacteria that can grow in fermented feeds. Alfalfa, a high-protein plant, is widely fed as hay; fescue is the least palatable and least nutritious hay. Mold in a goat's feed can make it sick and possibly kill it.

The digestive physiology of a very young kid (like the young of other ruminants) is essentially the same as that of a monogastric animal. Milk digestion begins in the abomasum, the milk having bypassed the rumen via closure of the reticuloesophageal groove during suckling. At birth, the rumen is undeveloped, but as the kid begins to consume solid feed, the rumen soon increases in size and in its capacity to absorb nutrients.

The adult size of a particular goat is a product of its breed (genetic potential) and its diet while growing (nutritional potential). As with all livestock, increased protein diets (10 to 14%) and sufficient calories during the prepuberty period yield higher growth rates and larger eventual size than lower protein rates and limited calories. Large-framed goats, with a greater skeletal size, reach mature weight at a later age (36 to 42 months) than small-framed goats (18 to 24 months) if both are fed to their full potential. Large-framed goats need more calories than small-framed goats for maintenance of daily functions.

Behavior

Goats are naturally curious. They are also agile and well known for their ability to climb and balance in precarious places. This makes them the only ruminant to regularly climb trees. Due to their agility and inquisitiveness, they are notorious for escaping their pens by testing fences and enclosures, either intentionally or simply because they are used to climb on. If any of the fencing can be overcome, goats will almost inevitably escape. Due to their intelligence, once a goat has discovered a weakness in the fence, they will exploit it repeatedly, and other goats will observe and quickly learn the same method.

Goats establish a dominance hierarchy in flocks, sometimes through head butting.

An example of the goats' social behavior within a flock.

Glycerinated goat tongue

Goats explore anything new or unfamiliar in their surroundings, primarily with their prehensile upper lip and tongue, by nibbling at them, occasionally even eating them.

When handled as a group, goats tend to display less herding behavior than sheep. When grazing undisturbed, they tend to spread across the field or range, rather than feed side-by-side as do sheep. When nursing young, goats will leave their kids separated ("lying out") rather than clumped, as do sheep. They will generally turn and face an intruder and bucks are more likely to charge or butt at humans than are rams.

A study by Queen Mary University reports that goats try to communicate with people in the same manner as domesticated animals such as dogs and horses. Goats were first domesticated as livestock more than 10,000 years ago. Research conducted to test communication skills found that the goats will look to a human for assistance when faced with a challenge that had previously been mastered, but was then modified. Specifically, when presented with a box, the goat was able to remove the lid and retrieve a treat inside, but when the box was turned so the lid could not be removed, the goat would turn and gaze at the person and move toward them, before looking back toward the box. This is the same type of complex communication observed by animals bred as domestic pets, such as dogs. Researchers believe that better understanding of human-goat interaction could offer overall improvement in the animals' welfare. The field of Anthrozoology has established that domesticated animals have the capacity for complex communication with humans when in 2015 a Japanese scientist determined that levels of oxytocin did increase in human subjects when dogs were exposed to a dose of the "love hormone", proving that the Human Animal Bond does exist. This is the same affinity that was proven with the London study above; goats are intelligent, capable of complex communication, and able to form bonds. Despite having the reputation of being slightly rebellious, more and more people today are choosing more exotic companion animals like goats. Goats are herd animals and typically prefer the company of other goats, but because of their herd mentality, they will follow their human around just the same.

Goats are well known for being hard to contain with fencing.

Diseases

While goats are generally considered hardy animals and in many situations receive little medical care, they are subject to a number of diseases. Among the conditions affecting goats are respiratory diseases including pneumonia, foot rot, internal parasites, pregnancy toxosis and feed toxicity. Feed toxicity can vary based on breed and location. Certain foreign fruits and vegetables can be toxic to different breeds of goats.

Goats can become infected with various viral and bacterial diseases, such as foot-and-mouth disease, caprine arthritis encephalitis, caseous lymphadenitis, pinkeye, mastitis, and pseudorabies. They can transmit a number of zoonotic diseases to people, such as tuberculosis, brucellosis, Q-fever, and rabies.

Life Expectancy

Life expectancy for goats is between fifteen and eighteen years. An instance of a goat reaching the age of 24 has been reported.

Several factors can reduce this average expectancy; problems during kidding can lower a doe's expected life span to ten or eleven, and stresses of going into rut can lower a buck's expected life span to eight to ten years.

Agriculture

A goat is useful to humans when it is living and when it is dead, first as a renewable provider of milk, manure, and fiber, and then as meat and hide. Some charities provide goats to impoverished people in poor countries, because goats are easier and cheaper to manage than cattle, and have multiple uses. In addition, goats are used for driving and packing purposes.

Goat husbandry is common through the Norte Chico region in Chile. Intensive goat husbrandry in drylands may produce severe erosion and desertification. Image from upper Limarí River

The intestine of goats is used to make "catgut", which is still in use as a material for internal human surgical sutures and strings for musical instruments. The horn of the goat, which signifies plenty and wellbeing (the cornucopia), is also used to make spoons.

The Boer goat – in this case a buck – is a widely kept meat breed.

Worldwide Goat Population Statistics

According to the Food and Agriculture Organization (FAO), the top producers of goat milk in 2008 were India (4 million metric tons), Bangladesh (2.16 million metric tons) and the Sudan (1.47 million metric tons).

Husbandry

Husbandry, or animal care and use, varies by region and culture. The particular housing used for goats depends not only on the intended use of the goat, but also on the region of the world where they are raised. Historically, domestic goats were generally kept in herds that wandered on hills or other grazing areas, often tended by goatherds who were frequently children or adolescents, similar to the more widely known shepherd. These methods of herding are still used today.

Reared goat(Husbandry)

In some parts of the world, especially Europe and North America, distinct breeds of goats are kept for dairy (milk) and for meat production. Excess male kids of dairy breeds are typically slaughtered for meat. Both does and bucks of meat breeds may be slaughtered for meat, as well as older animals of any breed. The meat of older bucks (more than one year old) is generally considered not desirable for meat for human consumption. Castration at a young age prevents the development of typical buck odor.

For smallholder farmers in many countries, such as this woman from Burkina Faso, goats are important livestock.

Dairy goats are generally pastured in summer and may be stabled during the winter. As dairy does are milked daily, they are generally kept close to the milking shed. Their grazing is typically supplemented with hay and concentrates. Stabled goats may be kept in stalls similar to horses, or in larger group pens. In the US system, does are generally rebred annually. In some European commercial dairy systems, the does are bred only twice, and are milked continuously for several years after the second kidding.

Meat goats are more frequently pastured year-round, and may be kept many miles from barns. Angora and other fiber breeds are also kept on pasture or range. Range-kept and pastured goats may be supplemented with hay or concentrates, most frequently during the winter or dry seasons.

In India, Nepal, and much of Asia, goats are kept largely for milk production, both in commercial and household settings. The goats in this area may be kept closely housed or may be allowed to range for fodder. The Salem Black goat is herded to pasture in fields and along roads during the day, but is kept penned at night for safe-keeping.

In Africa and the Mideast, goats are typically run in flocks with sheep. This maximizes the production per acre, as goats and sheep prefer different food plants. Multiple types of goat-raising are found in Ethiopia, where four main types have been identified: pastured in annual crop systems, in perennial crop systems, with cattle, and in arid areas, under pastoral (nomadic) herding systems. In all four systems, however, goats were typically kept in extensive systems, with few purchased inputs. Household goats are traditionally kept in Nigeria. While many goats are allowed to wander the homestead or village, others are kept penned and fed in what is called a 'cut-and-carry' system. This type of husbandry is also used in parts of Latin America. Cut-and-carry, which refers to the practice of cutting down grasses, corn or cane for feed rather than allowing the animal access to the field, is particularly suited for types of feed, such as corn or cane, that are easily destroyed by trampling.

Pet goats may be found in many parts of the world when a family keeps one or more animals for emotional reasons rather than as production animals. It is becoming more common for goats to be kept exclusively as pets in North America and Europe.

Meat

The taste of goat kid meat is similar to that of spring lamb meat; in fact, in the English-speaking islands of the Caribbean, and in some parts of Asia, particularly Bangladesh, Pakistan and India, the word "mutton" is used to describe both goat and lamb meat. However, some compare the taste of goat meat to veal or venison, depending on the age and condition of the goat. Its flavor is said to be primarily linked to the presence of 4-methyloctanoic and 4-methylnonanoic acid. It can be prepared in a variety of ways, including stewing, baking, grilling, barbecuing, canning, and frying; it can be minced, curried, or made into sausage. Due to its low fat content, the meat can toughen at high temperatures if cooked without additional moisture. One of the most popular goats grown for meat is the South African Boer, introduced into the United States in the early 1990s. The New Zealand Kiko is also considered a meat breed, as is the myotonic or "fainting goat", a breed originating in Tennessee.

Milk, Butter and Cheese

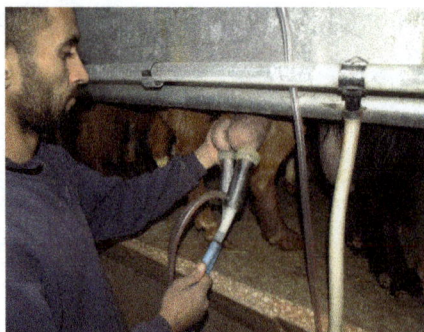

A goat being machine milked on an organic farm

Goats produce about 2% of the world's total annual milk supply. Some goats are bred specifically for milk. If the strong-smelling buck is not separated from the does, his scent will affect the milk.

Goat milk naturally has small, well-emulsified fat globules, which means the cream remains suspended in the milk, instead of rising to the top, as in raw cow milk; therefore, it does not need to be homogenized. Indeed, if the milk is to be used to make cheese, homogenization is not recommended, as this changes the structure of the milk, affecting the culture's ability to coagulate the milk and the final quality and yield of cheese.

Dairy goats in their prime (generally around the third or fourth lactation cycle) average—2.7 to 3.6 kg (6 to 8 lb)—of milk production daily—roughly 2.8 to 3.8 l (3 to 4 U.S. qt)—during a ten-month lactation, producing more just after freshening and gradually dropping in production toward the end of their lactation. The milk generally averages 3.5% butterfat.

Goat milk is commonly processed into cheese, butter, ice cream, yogurt, *cajeta* and other products. Goat cheese is known as *fromage de chèvre* ("goat cheese") in France. Some varieties include Rocamadour and Montrachet. Goat butter is white because goats produce milk with the yellow beta-carotene converted to a colorless form of vitamin A.

Nutrition

The American Academy of Pediatrics discourages feeding infants milk derived from goats. An April 2010 case report summarizes their recommendation and presents "a comprehensive review of the consequences associated with this dangerous practice", also stating, "Many infants are exclusively fed unmodified goat's milk as a result of cultural beliefs as well as exposure to false online information. Anecdotal reports have described a host of morbidities associated with that practice, including severe electrolyte abnormalities, metabolic acidosis, megaloblastic anemia, allergic reactions including life-threatening anaphylactic shock, hemolytic uremic syndrome, and infections." Untreated caprine brucellosis results in a 2% case fatality rate. According to the USDA, doe milk is not recommended for human infants because it contains "inadequate quantities of iron, folate, vitamins C and D, thiamine, niacin, vitamin B_6, and pantothenic acid to meet an infant's nutritional needs" and may cause harm to an infant's kidneys and could cause metabolic damage.

The Department of Health in the United Kingdom has repeatedly released statements stating on various occasions that "Goats' milk is not suitable for babies, and infant formulas and follow-on formulas based on goats' milk protein have not been approved for use in Europe", and "infant milks based on goats' milk protein are not suitable as a source of nutrition for infants."

Also according to the Canadian Federal Health Department – Health Canada, most of the dangers or counter-indication of feeding unmodified goat milk to infants, are similar to those incurring in the same practice with cow's milk, namely in the allergic reactions.

On the other hand, some farming groups promote the practice. For example, Small Farm Today in 2005 claimed beneficial use in invalid and convalescent diets, proposing that glycerol ethers, possibly important in nutrition for nursing infants, are much higher in doe milk than in cow milk. A 1970 book on animal breeding claimed doe milk differs from cow or human milk by having higher digestibility, distinct alkalinity, higher buffering capacity, and certain therapeutic values in human medicine and nutrition. George Mateljan suggested doe milk can replace ewe milk or cow milk in diets of those who are allergic to certain mammals' milk. However, like cow milk, doe milk has lactose (sugar), and may cause gastrointestinal problems for individuals with lactose intolerance. In fact, the level of lactose is similar to that of bovine milk.

Basic composition of various milks (mean values per 100 g)			
Constituent	**Doe (Goat)**	**Cow**	**Human**
Fat (g)	3.8	3.6	4.0
Protein (g)	3.5	3.3	1.2
Lactose (g)	4.1	4.6	6.9
Ash (g)	0.8	0.7	0.2
Total solids (g)	12.2	12.3	12.3
Calories	70	69	68

Milk composition analysis, per 100 grams					
Constituents	**unit**	**Cow**	**Doe (Goat)**	**Ewe (Sheep)**	**Water buffalo**
Water	g	87.8	88.9	83.0	81.1
Protein	g	3.2	3.1	5.4	4.5
Fat	g	3.9	3.5	6.0	8.0
Carbohydrate	g	4.8	4.4	5.1	4.9
Energy	kcal	66	60	95	110
Energy	kJ	275	253	396	463
Sugars (lactose)	g	4.8	4.4	5.1	4.9
Cholesterol	mg	14	10	11	8
Calcium	IU	120	100	170	195
Saturated fatty acids	g	2.4	2.3	3.8	4.2
Monounsaturated fatty acids	g	1.1	0.8	1.5	1.7
Polyunsaturated fatty acids	g	0.1	0.1	0.3	0.2

These compositions vary by breed (especially in the Nigerian Dwarf breed), animal, and point in the lactation period.

Fiber

The Angora breed of goats produces long, curling, lustrous locks of mohair. The entire body of the goat is covered with mohair and there are no guard hairs. The locks constantly grow to four inches or more in length. Angora crossbreeds, such as the pygora and the nigora, have been created to produce mohair and/or cashgora on a smaller, easier-to-manage animal. The wool is shorn twice a year, with an average yield of about 4.5 kg (10 lb).

An Angora goat

Most goats have softer insulating hairs nearer the skin, and longer guard hairs on the surface. The desirable fiber for the textile industry is the former, and it goes by several names (down, cashmere and pashmina). The coarse guard hairs are of little value as they are too coarse, difficult to spin and difficult to dye. The cashmere goat produces a commercial quantity of cashmere wool, which is one of the most expensive natural fibers commercially produced; cashmere is very fine and soft. The cashmere goat fiber is harvested once a year, yielding around 260 g (9 oz) of down.

In South Asia, cashmere is called "pashmina" (from Persian *pashmina*, "fine wool"). In the 18th and early 19th centuries, Kashmir (then called Cashmere by the British), had a thriving industry producing shawls from goat-hair imported from Tibet and Tartary through Ladakh. The shawls were introduced into Western Europe when the General in Chief of the French campaign in Egypt (1799–1802) sent one to Paris. Since these shawls were produced in the upper Kashmir and Ladakh region, the wool came to be known as "cashmere".

Land Clearing

Goats have been used by humans to clear unwanted vegetation for centuries. They have been described as "eating machines" and "biological control agents". There has been a resurgence of this in North America since 1990, when herds were used to clear dry brush from California hillsides thought to be endangered by potential wildfires. This form of using goats to clear land is sometimes known as conservation grazing. Since then, numerous public and private agencies have hired private herds to perform similar tasks. This practice has become popular in the Pacific Northwest, where they are used

to remove invasive species not easily removed by humans, including (thorned) black-berry vines and poison oak.

Use for Medical Training

As a goat's anatomy and physiology is not too dissimilar from that of human, some coun-tries' militaries use goats to train combat medics. In the United States, goats have become the main animal species used for this purpose after Pentagon phased out using dogs for medical training in the 1980s. While modern mannequins used in medical training are quite efficient in simulating the behavior of a human body, trainees feel that "the goat exercise provide[s] a sense of urgency that only real life trauma can provide".

As Pets

Some people choose goats as a pet because of their ability to form close bonds with their human guardians. Because of goats' herd mentality, they will follow their owner around and form close bonds with them.

Breeds

Goat breeds fall into overlapping, general categories. They are generally distributed in those used for dairy, fiber, meat, skins, and as companion animals. Some breeds are also particularly noted as pack goats.

Showing

Goat breeders' clubs frequently hold shows, where goats are judged on traits relat-ing to conformation, udder quality, evidence of high production, longevity, build and muscling (meat goats and pet goats) and fiber production and the fiber itself (fiber goats). People who show their goats usually keep registered stock and the offspring of award-winning animals command a higher price. Registered goats, in general, are usually higher-priced if for no other reason than that records have been kept proving their ancestry and the production and other data of their sires, dams, and other ances-tors. A registered doe is usually less of a gamble than buying a doe at random (as at an auction or sale barn) because of these records and the reputation of the breeder. Chil-dren's clubs such as 4-H also allow goats to be shown. Children's shows often include a showmanship class, where the cleanliness and presentation of both the animal and the exhibitor as well as the handler's ability and skill in handling the goat are scored. In a showmanship class, conformation is irrelevant since this is not what is being judged.

Various "Dairy Goat Scorecards" (milking does) are systems used for judging shows in the US. The American Dairy Goat Association (ADGA) scorecard for an adult doe includes a point system of a hundred total with major categories that include gener-al appearance, the dairy character of a doe (physical traits that aid and increase milk production), body capacity, and specifically for the mammary system. Young stock and bucks are judged by different scorecards which place more emphasis on the other three categories; general appearance, body capacity, and dairy character.

A Nigerian Dwarf milker in show clip. This doe is angular and dairy with a capacious and well supported mammary system.

The American Goat Society (AGS) has a similar, but not identical scorecard that is used in their shows. The miniature dairy goats may be judged by either of the two scorecards. The "Angora Goat scorecard" used by the Colored Angora Goat Breeder's Association (CAGBA), which covers the white and the colored goats, includes evaluation of an animal's fleece color, density, uniformity, fineness, and general body confirmation. Disqualifications include: a deformed mouth, broken down pasterns, deformed feet, crooked legs, abnormalities of testicles, missing testicles, more than 3 inch split in scrotum, and close-set or distorted horns.

Religion, Mythology and Folklore

According to Norse mythology, the god of thunder, Thor, has a chariot that is pulled by the goats Tanngrisnir and Tanngnjóstr. At night when he sets up camp, Thor eats the meat of the goats, but takes care that all bones remain whole. Then he wraps the remains up, and in the morning, the goats always come back to life to pull the chariot. When a farmer's son who is invited to share the meal breaks one of the goats' leg bones to suck the marrow, the animal's leg remains broken in the morning, and the boy is forced to serve Thor as a servant to compensate for the damage.

An ancient Greek *oenochoe* depicting wild goats

Possibly related, the Yule Goat is one of the oldest Scandinavian and Northern European Yule and Christmas symbols and traditions. Yule Goat originally denoted the goat that was slaughtered around Yule, but it may also indicate a goat figure made out of straw. It is also used about the custom of going door-to-door singing carols and getting food and drinks in return, often fruit, cakes and sweets. "Going Yule Goat" is similar to the British custom wassailing, both with heathen roots. The Gävle Goat is a giant version of the Yule Goat, erected every year in the Swedish city of Gävle.

The Greek god Pan is said to have the upper body of a man and the horns and lower body of a goat. Pan was a very lustful god, nearly all of the myths involving him had to do with him chasing nymphs. He is also credited with creating the pan flute.

The goat is one of the twelve-year cycle of animals which appear in the Chinese zodiac related to the Chinese calendar. Each animal is associated with certain personality traits; those born in a year of the goat are predicted to be shy, introverted, creative, and perfectionist.

Amalthée et la chèvre de Jupiter (Amalthea and Jupiter's goat); commissioned by the Queen of France in 1787 for the royal dairy at Rambouillet

Several mythological hybrid creatures are believed to consist of parts of the goat, including the Chimera. The Capricorn sign in the Western zodiac is usually depicted as a goat with a fish's tail. Fauns and satyrs are mythological creatures that are part goat and part human. The mineral bromine is named from the Greek word "bromos", which means "stench of he-goats".

Goats are mentioned many times in the Bible. A goat is considered a "clean" animal by Jewish dietary laws and was slaughtered for an honored guest. It was also acceptable for some kinds of sacrifices. Goat-hair curtains were used in the tent that contained the tabernacle (Exodus 25:4). Its horns can be used instead of sheep's horn to make a shofar. On Yom Kippur, the festival of the Day of Atonement, two goats were chosen and lots were drawn for them. One was sacrificed and the other allowed to escape into the wilderness, symbolically carrying with it the sins of the community. From this comes the word "scapegoat". A leader or king was sometimes compared to a male goat leading the flock. In the New Testament, Jesus told a parable of the Sheep and the Goats (Gospel of Matthew 25).

Popular Christian folk tradition in Europe associated Satan with imagery of goats. A common superstition in the Middle Ages was that goats whispered lewd sentences in the ears of the saints. The origin of this belief was probably the behavior of the buck in rut, the very epitome of lust. The common medieval depiction of the Devil was that of a goat-like face with horns and small beard (a goatee). The Black Mass, a probably mythological "Satanic mass", was said to involve a black goat, the form in which Satan supposedly manifested himself for worship.

The goat has had a lingering connection with Satanism and pagan religions, even into modern times. The inverted pentagram, a symbol used in Satanism, is said to be shaped like a goat's head. The "Baphomet of Mendes" refers to a satanic goat-like figure from 19th-century occultism.

The common Russian surname *Kozlov* (Russian: Козло́в), means "goat". Goatee refers to a style of facial hair incorporating hair on a man's chin, so named because of some similarity to a goat's facial feature.

Feral Goats

Goats readily revert to the wild (become feral) if given the opportunity. The only domestic animal known to return to feral life as swiftly is the cat. Feral goats have established themselves in many areas: they occur in Australia, New Zealand, Great Britain, the Galapagos and in many other places. When feral goats reach large populations in habitats which provide unlimited water supply and which do not contain sufficient large predators or which are otherwise vulnerable to goats' aggressive grazing habits, they may have serious effects, such as removing native scrub, trees and other vegetation which is required by a wide range of other creatures, not just other grazing or browsing animals. Feral goats are common in Australia. However, in other circumstances where predator pressure is maintained, they may be accommodated into some balance in the local food web.

Feral goat in Aruba

Sheep

The sheep (*Ovis aries*) is a quadrupedal, ruminant mammal typically kept as livestock. Like all ruminants, sheep are members of the order Artiodactyla, the even-toed ungulates. Although the name "sheep" applies to many species in the genus *Ovis*, in everyday usage it almost always refers to *Ovis aries*. Numbering a little over one billion, domestic sheep are also the most numerous species of sheep. An adult female sheep is referred to as a *ewe* (/ju□/), an intact male as a *ram* or occasionally a *tup*, a castrated male as a *wether*, and a younger sheep as a *lamb*.

Sheep are most likely descended from the wild mouflon of Europe and Asia. One of the earliest animals to be domesticated for agricultural purposes, sheep are raised for fleece, meat (lamb, hogget or mutton) and milk. A sheep's wool is the most widely used animal fiber, and is usually harvested by shearing. Ovine meat is called lamb when from younger animals and mutton when from older ones. Sheep continue to be important for wool and meat today, and are also occasionally raised for pelts, as dairy animals, or as model organisms for science.

Sheep husbandry is practised throughout the majority of the inhabited world, and has been fundamental to many civilizations. In the modern era, Australia, New Zealand, the southern and central South American nations, and the British Isles are most closely associated with sheep production.

Sheepraising has a large lexicon of unique terms which vary considerably by region and dialect. Use of the word *sheep* began in Middle English as a derivation of the Old English word *scēap*; it is both the singular and plural name for the animal. A group of sheep is called a flock, herd or mob. Many other specific terms for the various life stages of sheep exist, generally related to lambing, shearing, and age.

Being a key animal in the history of farming, sheep have a deeply entrenched place in human culture, and find representation in much modern language and symbology. As livestock, sheep are most often associated with pastoral, Arcadian imagery. Sheep figure in many mythologies—such as the Golden Fleece—and major religions, especially the Abrahamic traditions. In both ancient and modern religious ritual, sheep are used as sacrificial animals.

Description and evolution

Domestic sheep are relatively small ruminants, usually with a crimped hair called wool and often with horns forming a lateral spiral. Domestic sheep differ from their wild relatives and ancestors in several respects, having become uniquely neotenic as a result of selective breeding by humans. A few primitive breeds of sheep retain some of the characteristics of their wild cousins, such as short tails. Depending on breed, domestic sheep may have no horns at all (i.e. polled), or horns in both sexes, or in males only. Most horned breeds have a single pair, but a few breeds may have several.

Another trait unique to domestic sheep as compared to wild ovines is their wide variation in color. Wild sheep are largely variations of brown hues, and variation within species is extremely limited. Colors of domestic sheep range from pure white to dark chocolate brown, and even spotted or piebald. Selection for easily dyeable white fleeces began early in sheep domestication, and as white wool is a dominant trait it spread quickly. However, colored sheep do appear in many modern breeds, and may even appear as a recessive trait in white flocks. While white wool is desirable for large commercial markets, there is a niche market for colored fleeces, mostly for handspinning. The nature of the fleece varies widely among the breeds, from dense and highly crimped, to long and hairlike. There is variation of wool type and quality even among members of the same flock, so wool classing is a step in the commercial processing of the fibre.

Suffolks are a medium wool, black-faced breed of meat sheep that make up 60% of the sheep population in the U.S.

Depending on breed, sheep show a range of heights and weights. Their rate of growth and mature weight is a heritable trait that is often selected for in breeding. Ewes typically weigh between 45 and 100 kilograms (100 and 220 lb), and rams between 45 and 160 kilograms (100 and 350 lb). When all deciduous teeth have erupted, the sheep has 20 teeth. Mature sheep have 32 teeth. As with other ruminants, the front teeth in the lower jaw bite against a hard, toothless pad in the upper jaw. These are used to pick off vegetation, then the rear teeth grind it before it is swallowed. There are eight lower front teeth in ruminants, but there is some disagreement as to whether these are eight incisors, or six incisors and two incisor-shaped canines. This means that the dental formula for sheep is either 0.0.3.34.0.3.3 or 0.0.3.33.1.3.3 There is a large diastema between the incisors and the molars. In the first few years of life one can calculate the age of sheep from their front teeth, as a pair of milk teeth is replaced by larger adult teeth each year, the full set of eight adult front teeth being complete at about four years of age. The front teeth are then gradually lost as sheep age, making it harder for them to feed and hindering the health and productivity of the animal. For this reason, domestic sheep on normal pasture begin to slowly decline from four years on, and the life expectancy of a sheep is 10 to 12 years, though some sheep may live as long as 20 years.

Sheep have good hearing, and are sensitive to noise when being handled. Sheep have horizontal slit-shaped pupils, with excellent peripheral vision; with visual fields of about 270° to 320°, sheep can see behind themselves without turning their heads. Many breeds have only short hair on the face, and some have facial wool (if any) confined to the poll and or the area of the mandibular angle; the wide angles of peripheral vision apply to these breeds. A few breeds tend to have considerable wool on the face; for some individuals of these breeds, peripheral vision may be greatly reduced by "wool blindness", unless recently shorn about the face. Sheep have poor depth perception; shadows and dips in the ground may cause sheep to baulk. In general, sheep have a tendency to move out of the dark and into well-lit areas, and prefer to move uphill when disturbed. Sheep also have an excellent sense of smell, and, like all species of their genus, have scent glands just in front of the eyes, and interdigitally on the feet. The purpose of these glands is uncertain, but those on the face may be used in breeding behaviors. The foot glands might also be related to reproduction, but alternative reasons, such as secretion of a waste product or a scent marker to help lost sheep find their flock, have also been proposed.

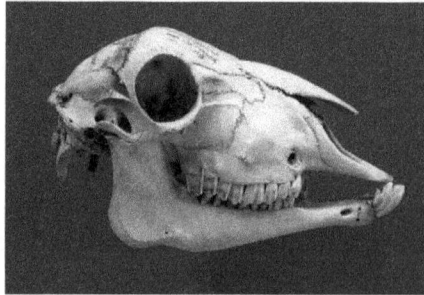

Skull

Sheep Compared to Goats

Sheep and goats are closely related: both are in the subfamily Caprinae. However, they are separate species, so hybrids rarely occur, and are always infertile. A hybrid of a ewe and a buck (a male goat) is called a sheep-goat hybrid (only a single such animal has been confirmed), and is not to be confused with the sheep-goat chimera, though both are known as "geep". Visual differences between sheep and goats include the beard of goats and divided upper lip of sheep. Sheep tails also hang down, even when short or docked, while the short tails of goats are held upwards. Sheep breeds are also often naturally polled (either in both sexes or just in the female), while naturally polled goats are rare (though many are polled artificially). Males of the two species differ in that buck goats acquire a unique and strong odor during the rut, whereas rams do not.

Breeds

The domestic sheep is a multi-purpose animal, and the more than 200 breeds now in existence were created to serve these diverse purposes. Some sources give a count of a thousand or more breeds, but these numbers cannot be verified, according to some sources. However, several hundred breeds of sheep have been identified by the FAO

(Food and Agriculture Organization of the UN), with the estimated number varying somewhat from time to time: e.g. 863 breeds as of 1993, 1314 breeds as of 1995 and 1229 breeds as of 2006. (These numbers exclude extinct breeds, which are also tallied by the FAO.) For the purpose of such tallies, the FAO definition of a breed is "either a subspecific group of domestic livestock with definable and identifiable external characteristics that enable it to be separated by visual appraisal from other similarly defined groups within the same species or a group for which geographical and/or cultural separation from phenotypically similar groups has led to acceptance of its separate identity." Almost all sheep are classified as being best suited to furnishing a certain product: wool, meat, milk, hides, or a combination in a dual-purpose breed. Other features used when classifying sheep include face color (generally white or black), tail length, presence or lack of horns, and the topography for which the breed has been developed. This last point is especially stressed in the UK, where breeds are described as either upland (hill or mountain) or lowland breeds. A sheep may also be of a fat-tailed type, which is a dual-purpose sheep common in Africa and Asia with larger deposits of fat within and around its tail.

Sheep being judged for adherence to their breed standard, and being held by the most common method of restraint

The Barbados Blackbelly is a hair sheep breed of Caribbean origin.

Breeds are often categorized by the type of their wool. Fine wool breeds are those that have wool of great crimp and density, which are preferred for textiles. Most of these were derived from Merino sheep, and the breed continues to dominate the world sheep industry. Downs breeds have wool between the extremes, and are typically fast-growing meat and ram breeds with dark faces. Some major medium wool breeds, such as the Corriedale, are dual-purpose crosses of long and fine-wooled breeds and were created for high-production commercial flocks. Long wool breeds are the largest of sheep, with long wool and a slow rate of growth. Long wool sheep are most valued for crossbreeding to improve the attributes of other sheep types. For example: the American Columbia breed was developed by crossing Lincoln rams (a long wool breed) with fine-wooled Rambouillet ewes.

Coarse or carpet wool sheep are those with a medium to long length wool of characteristic coarseness. Breeds traditionally used for carpet wool show great variability, but the chief requirement is a wool that will not break down under heavy use (as would that of the finer breeds). As the demand for carpet-quality wool declines, some breeders of this type of sheep are attempting to use a few of these traditional breeds for alternative purposes. Others have always been primarily meat-class sheep.

A minor class of sheep are the dairy breeds. Dual-purpose breeds that may primarily be meat or wool sheep are often used secondarily as milking animals, but there are a few breeds that are predominantly used for milking. These sheep produce a higher quantity of milk and have slightly longer lactation curves. In the quality of their milk, the fat and protein content percentages of dairy sheep vary from non-dairy breeds, but lactose content does not.

The Lička pramenka is a sheep breed of Croatian origin

A last group of sheep breeds is that of fur or hair sheep, which do not grow wool at all. Hair sheep are similar to the early domesticated sheep kept before woolly breeds were developed, and are raised for meat and pelts. Some modern breeds of hair sheep, such as the Dorper, result from crosses between wool and hair breeds. For meat and hide producers, hair sheep are cheaper to keep, as they do not need shearing. Hair sheep are also more resistant to parasites and hot weather.

With the modern rise of corporate agribusiness and the decline of localized family farms, many breeds of sheep are in danger of extinction. The Rare Breeds Survival Trust of the UK lists 22 native breeds as having only 3,000 registered animals (each), and The Livestock Conservancy lists 14 as either "critical" or "threatened". Preferences for breeds with uniform characteristics and fast growth have pushed heritage (or heirloom) breeds to the margins of the sheep industry. Those that remain are maintained through the efforts of conservation organizations, breed registries, and individual farmers dedicated to their preservation.

Diet

Sheep are exclusively herbivorous mammals. Most breeds prefer to graze on grass and other short roughage, avoiding the taller woody parts of plants that goats readily consume. Both sheep and goats use their lips and tongues to select parts of the plant that are easier to digest or higher in nutrition. Sheep, however, graze well in monoculture pastures where most goats fare poorly. Like all ruminants, sheep have a complex digestive system composed of four chambers, allowing them to break down cellulose from stems, leaves, and seed hulls into simpler carbohydrates. When sheep graze, vegetation is chewed into a mass called a bolus, which is then passed into the rumen, via the reticulum. The rumen is a 19- to 38-liter (5 to 10 gal) organ in which feed is fermented. The fermenting organisms include bacteria, fungi, and protozoa. (Other important rumen organisms include some archaea, which produce methane from carbon dioxide.) The bolus is periodically regurgitated back to the mouth as cud for additional chewing and salivation. Cud chewing is an adaptation allowing ruminants to graze more quickly in the morning, and then fully chew and digest feed later in the day. This is safer than grazing, which requires lowering the head thus leaving the animal vulnerable to predators, while cud chewing does not.

Head of polled, domesticated sheep in the long grass

Other than forage, the other staple feed for sheep is hay, often during the winter months. The ability to thrive solely on pasture (even without hay) varies with breed, but all sheep can survive on this diet. Also included in some sheep's diets are minerals, either in a trace mix or in licks.

Grazing Behavior

Sheep follow a diurnal pattern of activity, feeding from dawn to dusk, stopping sporadically to rest and chew their cud. Ideal pasture for sheep is not lawnlike grass, but an array of grasses, legumes and forbs. Types of land where sheep are raised vary widely, from pastures that are seeded and improved intentionally to rough, native lands. Common plants toxic to sheep are present in most of the world, and include (but are not limited to) cherry, some oaks and acorns, tomato, yew, rhubarb, potato, and rhododendron.

Sheep grazing on public land

Effects on Pasture

Sheep are largely grazing herbivores, unlike browsing animals such as goats and deer that prefer taller foliage. With a much narrower face, sheep crop plants very close to the ground and can overgraze a pasture much faster than cattle. For this reason, many shepherds use managed intensive rotational grazing, where a flock is rotated through multiple pastures, giving plants time to recover. Paradoxically, sheep can both cause and solve the spread of invasive plant species. By disturbing the natural state of pasture, sheep and other livestock can pave the way for invasive plants. However, sheep also prefer to eat invasives such as cheatgrass, leafy spurge, kudzu and spotted knapweed over native species such as sagebrush, making grazing sheep effective for conservation grazing. Research conducted in Imperial County, California compared lamb grazing with herbicides for weed control in seedling alfalfa fields. Three trials demonstrated that grazing lambs were just as effective as herbicides in controlling winter weeds. Entomologists also compared grazing lambs to insecticides for insect control in winter alfalfa. In this trial, lambs provided insect control as effectively as insecticides.

Rumination

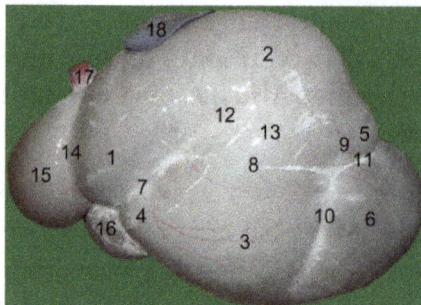
Ruminant system

During fermentation, the rumen produces gas that must be expelled; disturbances of the organ, such as sudden changes in a sheep's diet, can cause the potentially fatal condition of bloat, when gas becomes trapped in the rumen, due to reflex closure of the caudal esophageal sphincter when in contact with foam or liquid. After fermentation in the rumen, feed passes into the reticulum and the omasum; special feeds such as grains may bypass the rumen altogether. After the first three chambers, food moves into the abomasum for final digestion before processing by the intestines. The abomasum is the only one of the four chambers analogous to the human stomach, and is sometimes called the "true stomach".

Concentrated Diets

Sheep are one of the few livestock animals raised for meat today that have rarely been raised in an intensive, confined animal feeding operation (CAFO). Although there is a growing movement advocating alternative farming styles, a large percentage of beef cattle, pigs, and poultry are still produced under such conditions. In contrast, only some sheep are regularly given high-concentration grain feed, much less kept in confinement. Especially in industrialized countries, sheep producers may fatten market lambs before slaughter (called "finishing") in feedlots. Many sheep breeders flush ewes and rams with a daily ration of grain during breeding to increase fertility. Ewes may be flushed during pregnancy to increase birth weights, as 70% of a lamb's growth occurs in the last five to six weeks of gestation. However, overfeeding of ewe hoggets (i.e. adolescent ewes) in early pregnancy can result in restricted placental development, restricting growth of fetal lambs in late pregnancy. Otherwise, only lactating ewes and especially old or infirm sheep are commonly provided with grain. Feed provided to sheep must be specially formulated, as most cattle, poultry, pig, and even some goat feeds contain levels of copper that are lethal to sheep. The same danger applies to mineral supplements such as salt licks.

Behavior

A flock of sheep following a leader

Sheep showing flocking behavior during a sheepdog trial

Flock Behavior

Sheep are flock animals and strongly gregarious; much sheep behavior can be understood on the basis of these tendencies. The dominance hierarchy of sheep and their natural inclination to follow a leader to new pastures were the pivotal factors in sheep being one of the first domesticated livestock species. Furthermore, in contrast to the red deer and gazelle (two other ungulates of primary importance to meat production in prehistoric times), sheep do not defend territories although they do form home ranges. All sheep have a tendency to congregate close to other members of a flock, although this behavior varies with breed, and sheep can become stressed when separated from their flock members. During flocking, sheep have a strong tendency to follow and a leader may simply be the first individual to move. Relationships in flocks tend to be closest among related sheep: in mixed-breed flocks, subgroups of the same breed tend to form, and a ewe and her direct descendants often move as a unit within large flocks. Sheep can become hefted to one particular local pasture (heft) so they do not roam freely in unfenced landscapes. Lambs learn the heft from ewes and if whole flocks are culled it must be retaught to the replacement animals.

Flock behaviour in sheep is generally only exhibited in groups of four or more sheep; fewer sheep may not react as expected when alone or with few other sheep. Being a prey species, the primary defense mechanism of sheep is to flee from danger when their flight zone is entered. Cornered sheep may charge and butt, or threaten by hoof stamping and adopting an aggressive posture. This is particularly true for ewes with newborn lambs.

In regions where sheep have no natural predators, none of the native breeds of sheep exhibit a strong flocking behavior.

Herding

Escaped sheep being led back to pasture with the enticement of food. This method of moving sheep works best with smaller flocks.

Farmers exploit flocking behavior to keep sheep together on unfenced pastures such as hill farming, and to move them more easily. For this purpose shepherds may use herding dogs in this effort, with a highly bred herding ability. Sheep are food-oriented, and association of humans with regular feeding often results in sheep soliciting people for food. Those who are moving sheep may exploit this behavior by leading sheep with buckets of feed.

Dominance Hierarchy

Sheep establish a dominance hierarchy through fighting, threats and competitiveness. Dominant animals are inclined to be more aggressive with other sheep, and usually feed first at troughs. Primarily among rams, horn size is a factor in the flock hierarchy. Rams with different size horns may be less inclined to fight to establish the dominance order, while rams with similarly sized horns are more so. Merinos have an almost linear hierarchy whereas there is a less rigid structure in Border Leicesters when a competitive feeding situation arises.

In sheep, position in a moving flock is highly correlated with social dominance, but there is no definitive study to show consistent voluntary leadership by an individual sheep.

Intelligence and Learning Ability

Sheep are frequently thought of as unintelligent animals. Their flocking behavior and quickness to flee and panic can make shepherding a difficult endeavor for the uninitiated. Despite these perceptions, a University of Illinois monograph on sheep reported them to be just below pigs and on par with cattle in IQ. Sheep can recognize individual human and ovine faces, and remember them for years. In addition to long-term facial recognition of individuals, sheep can also differentiate emotional states through facial characteristics. If worked with patiently, sheep may learn their names and many sheep are trained to be led by halter for showing and other purposes. Sheep have also responded well to clicker training. Sheep have been used as pack animals; Tibetan nomads distribute baggage equally throughout a flock as it is herded between living sites.

It has been reported that some sheep have apparently shown problem-solving abilities; a flock in West Yorkshire, England allegedly found a way to get over cattle grids by rolling on their backs, although documentation of this has relied on anecdotal accounts.

Vocalisations

Sounds made by domestic sheep include bleats, grunts, rumbles and snorts. Bleating ("baaing") is used mostly for contact communication, especially between dam and lambs, but also at times between other flock members. The bleats of individual sheep are distinctive, enabling the ewe and her lambs to recognize each other's vocalizations. Vocal communication between lambs and their dam declines to a very low level within several weeks after parturition. A variety of bleats may be heard, depending on sheep age and circumstances. Apart from contact communication, bleating may signal distress, frustration or impatience; however, sheep are usually silent when in pain.

Isolation commonly prompts bleating by sheep. Pregnant ewes may grunt when in labor. Rumbling sounds are made by the ram during courting; somewhat similar rumbling sounds may be made by the ewe, especially when with her neonate lambs. A snort (explosive exhalation through the nostrils) may signal aggression or a warning, and is often elicited from startled sheep.

Senses

In sheep breeds lacking facial wool, the visual field is wide. In 10 sheep (Cambridge, Lleyn and Welsh Mountain breeds, which lack facial wool), the visual field ranged from 298° to 325°, averaging 313.1°, with binocular overlap ranging from 44.5° to 74°, averaging 61.7°. In some breeds, unshorn facial wool can limit the visual field; in some individuals, this may be enough to cause "wool blindness". In 60 Merinos, visual fields ranged from 219.1° to 303.0°, averaging 269.9°, and the binocular field ranged from 8.9° to 77.7°, averaging 47.5°; 36% of the measurements were limited by wool, although photographs of the experiments indicate that only limited facial wool regrowth had occurred since shearing. In addition to facial wool (in some breeds), visual field limitations can include ears and (in some breeds) horns, so the visual field can be extended by tilting the head. Sheep eyes exhibit very low hyperopia and little astigmatism. Such visual characteristics are likely to produce a well-focused retinal image of objects in both the middle and long distance. Because sheep eyes have no accommodation, one might expect the image of very near objects to be blurred, but a rather clear near image could be provided by the tapetum and large retinal image of the sheep's eye, and adequate close vision may occur at muzzle length. Good depth perception, inferred from the sheep's sure-footedness, was confirmed in "visual cliff" experiments; behavioral responses indicating depth perception are seen in lambs at one day old. Sheep are thought to have colour vision, and can distinguish between a variety of colours: black, red, brown, green, yellow and white. Sight is a vital part of sheep communication, and when grazing, they maintain visual contact with each other. Each sheep lifts its head upwards to check the position of other sheep in the flock. This constant monitoring is probably what keeps the sheep in a flock as they move along grazing. Sheep become stressed when isolated; this stress is reduced if they are provided with a mirror, indicating that the sight of other sheep reduces stress.

Lamb

Taste is the most important sense in sheep, establishing forage preferences, with sweet and sour plants being preferred and bitter plants being more commonly rejected. Touch and sight are also important in relation to specific plant characteristics, such as succulence and growth form.

The ram uses his vomeronasal organ (sometimes called the Jacobson's organ) to sense the pheromones of ewes and detect when they are in estrus. The ewe uses her vomeronasal organ for early recognition of her neonate lamb.

Reproduction

Sheep follow a similar reproductive strategy to other herd animals. A group of ewes is generally mated by a single ram, who has either been chosen by a breeder or (in feral populations) has established dominance through physical contest with other rams. Most sheep are seasonal breeders, although some are able to breed year-round. Ewes generally reach sexual maturity at six to eight months old, and rams generally at four to six months. However, there are exceptions. For example, Finnsheep ewe lambs may reach puberty as early as 3 to 4 months, and Merino ewes sometimes reach puberty at 18 to 20 months. Ewes have estrus cycles about every 17 days, during which they emit a scent and indicate readiness through physical displays towards rams. A minority of rams (8% on average) display a preference for homosexuality and a small number of the females that were accompanied by a male fetus *in utero* are freemartins (female animals that are behaviorally masculine and lack functioning ovaries).

The second of twins being born.

In feral sheep, rams may fight during the rut to determine which individuals may mate with ewes. Rams, especially unfamiliar ones, will also fight outside the breeding period to establish dominance; rams can kill one another if allowed to mix freely. During the rut, even usually friendly rams may become aggressive towards humans due to increases in their hormone levels.

After mating, sheep have a gestation period of about five months, and normal labor takes one to three hours. Although some breeds regularly throw larger litters of lambs, most produce single or twin lambs. During or soon after labor, ewes and lambs may be confined to small lambing jugs, small pens designed to aid both careful observation of ewes and to cement the bond between them and their lambs.

A lamb's first steps

Ovine obstetrics can be problematic. By selectively breeding ewes that produce multiple offspring with higher birth weights for generations, sheep producers have inadvertently caused some domestic sheep to have difficulty lambing; balancing ease of lambing with high productivity is one of the dilemmas of sheep breeding. In the case of any such problems, those present at lambing may assist the ewe by extracting or repositioning lambs. After the birth, ewes ideally break the amniotic sac (if it is not broken during labor), and begin licking clean the lamb. Most lambs will begin standing within an hour of birth. In normal situations, lambs nurse after standing, receiving vital colostrum milk. Lambs that either fail to nurse or are rejected by the ewe require help to survive, such as bottle-feeding or fostering by another ewe.

After lambs are several weeks old, lamb marking (ear tagging, docking, and castrating) is carried out. Vaccinations are usually carried out at this point as well. Ear tags with numbers are attached, or ear marks are applied, for ease of later identification of sheep. Castration is performed on ram lambs not intended for breeding, although some shepherds choose to omit this for ethical, economic or practical reasons. However, many would disagree with regard to timing. Docking and castration are commonly done after 24 hours (to avoid interference with maternal bonding and consumption of colostrum) and are often done not later than one week after birth, to minimize pain, stress, recovery time and complications. The first course of vaccinations (commonly anti-clostridial) is commonly given at an age of about 10 to 12 weeks; i.e. when the concentration of maternal antibodies passively acquired via colostrum is expected to have fallen low enough to permit development of active immunity. Ewes are often revaccinated annually about 3 weeks before lambing, to provide high antibody concentrations in colostrum during the first several hours after lambing. Ram lambs that will either be slaughtered or separated from ewes before sexual maturity are not usually castrated. Tail docking is commonly done for welfare, having been shown to reduce risk of fly strike. Objections to all these procedures have been raised by animal rights groups, but farmers defend them by saying they solve many practical and veterinary problems, and inflict only temporary pain.

Health

Sheep may fall victim to poisons, infectious diseases, and physical injuries. As a prey species, a sheep's system is adapted to hide the obvious signs of illness, to prevent being targeted by predators. However, some signs of ill health are obvious, with sick sheep eating little, vocalizing excessively, and being generally listless. Throughout history, much of the money and labor of sheep husbandry has aimed to prevent sheep ailments. Historically, shepherds often created remedies by experimentation on the farm. In some developed countries, including the United States, sheep lack the economic importance for drug companies to perform expensive clinical trials required to approve more than a relatively limited number of drugs for ovine use. However, extra-label drug use in sheep production is permitted in many jurisdictions, subject to certain restrictions. In the US, for example, regulations governing extra-label drug use in animals are found in 21 CFR (Code of Federal Regulations) Part 530. In the 20th and 21st centuries, a minority of sheep owners have turned to alternative treatments such as homeopathy, herbalism and even traditional Chinese medicine to treat sheep veterinary problems. Despite some favorable anecdotal evidence, the effectiveness of alternative veterinary medicine has been met with skepticism in scientific journals. The need for traditional anti-parasite drugs and antibiotics is widespread, and is the main impediment to certified organic farming with sheep.

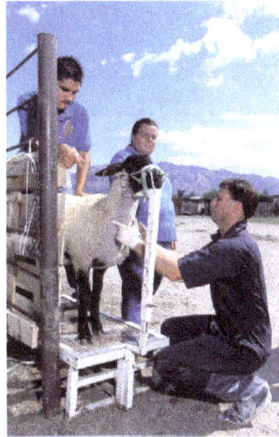

A veterinarian draws blood to test for resistance to scrapie

Many breeders take a variety of preventive measures to ward off problems. The first is to ensure all sheep are healthy when purchased. Many buyers avoid outlets known to be clearing houses for animals culled from healthy flocks as either sick or simply inferior. This can also mean maintaining a closed flock, and quarantining new sheep for a month. Two fundamental preventive programs are maintaining good nutrition and reducing stress in the sheep. Restraint, isolation, loud noises, novel situations, pain, heat, extreme cold, fatigue and other stressors can lead to secretion of cortisol, a stress hormone, in amounts that may indicate welfare problems. Excessive stress can compromise the immune system. "Shipping fever" (pneumonic mannheimiosis, formerly called pasteurellosis) is a disease of particular concern, that can occur as a result of stress, notably during transport and (or) handling. Pain, fear and several other stressors can cause secretion of epinephrine (adrenaline). Considerable epinephrine secretion in the final days before slaughter can adversely affect meat quality (by causing glycogenolysis, removing the substrate for normal post-slaughter acidification of meat) and result in meat becoming more susceptible to colonization by spoilage bacteria. Because of such issues, low-stress handling is essential in sheep management. Avoiding poisoning is also important; common poisons are pesticide sprays, inorganic fertilizer, motor oil, as well as radiator coolant containing ethylene glycol.

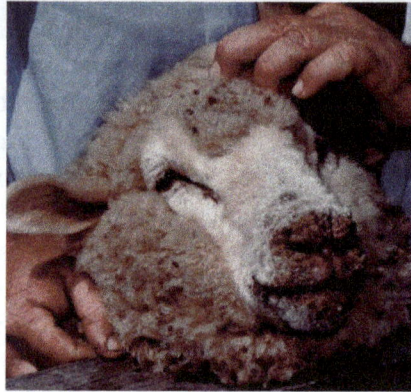

A sheep infected with orf, a disease transmittable to humans through skin contact

Common forms of preventive medication for sheep are vaccinations and treatments for parasites. Both external and internal parasites are the most prevalent malady in sheep, and are either fatal, or reduce the productivity of flocks. Worms are the most common internal parasites. They are ingested during grazing, incubate within the sheep, and are expelled through the digestive system (beginning the cycle again). Oral anti-parasitic medicines, known as drenches, are given to a flock to treat worms, sometimes after worm eggs in the feces has been counted to assess infestation levels. Afterwards, sheep may be moved to a new pasture to avoid ingesting the same parasites. External sheep parasites include: lice (for different parts of the body), sheep keds, nose bots, sheep itch mites, and maggots. Keds are blood-sucking parasites that cause general malnutrition and decreased productivity, but are not fatal. Maggots are those of the bot fly and the blow-fly. Fly maggots cause the extremely destructive condition of flystrike. Flies lay their eggs in wounds or wet, manure-soiled wool; when the maggots hatch they burrow into a sheep's flesh, eventually causing death if untreated. In addition to other treatments, crutching (shearing wool from a sheep's rump) is a common preventive method. Some countries allow mulesing, a practice that involves stripping away the skin on the rump to prevent fly-strike, normally performed when the sheep is a lamb. Nose bots are fly larvae that inhabit a sheep's sinuses, causing breathing difficulties and discomfort. Common signs are a discharge from the nasal passage, sneezing, and frantic movement such as head shaking. External parasites may be controlled through the use of backliners, sprays or immersive sheep dips.

A wide array of bacterial and viral diseases affect sheep. Diseases of the hoof, such as foot rot and foot scald may occur, and are treated with footbaths and other remedies. These painful conditions cause lameness and hinder feeding. Ovine Johne's disease is a wasting disease that affects young sheep. Bluetongue disease is an insect-borne illness causing fever and inflammation of the mucous membranes. Ovine rinderpest (or *peste des petits ruminants*) is a highly contagious and often fatal viral disease affecting sheep and goats.

A few sheep conditions are transmissible to humans. Orf (also known as scabby mouth, contagious ecthyma or soremouth) is a skin disease leaving lesions that is transmitted through skin-to-skin contact. Cutaneous anthrax is also called woolsorter's disease, as the spores can be transmitted in unwashed wool. More seriously, the organisms that can cause spontaneous enzootic abortion in sheep are easily transmitted to pregnant women. Also of concern are the prion disease scrapie and the virus that causes foot-and-mouth disease (FMD), as both can devastate flocks. The latter poses a slight risk to humans. During the 2001 FMD pandemic in the UK, hundreds of sheep were culled and some rare British breeds were at risk of extinction due to this.

Predation

Other than parasites and disease, predation is a threat to sheep and the profitability of sheep raising. Sheep have little ability to defend themselves, compared with other species kept as livestock. Even if sheep survive an attack, they may die from their injuries, or simply from panic. However, the impact of predation varies dramatically with region. In Africa, Australia, the Americas, and parts of Europe and Asia predators are a serious problem. In the United States, for instance, over one third of sheep deaths in 2004 were caused by predation. In contrast, other nations are virtually devoid of sheep predators, particularly islands known for extensive sheep husbandry. Worldwide, canids—including the domestic dog—are responsible for most sheep deaths. Other animals that occasionally prey on sheep include: felines, bears, birds of prey, ravens and feral hogs.

A lamb being attacked by coyotes with a bite to the throat

Sheep producers have used a wide variety of measures to combat predation. Pre-modern shepherds used their own presence, livestock guardian dogs, and protective structures such as barns and fencing. Fencing (both regular and electric), penning sheep at night and lambing indoors all continue to be widely used. More modern shepherds used guns, traps, and poisons to kill predators, causing significant decreases in predator populations. In the wake of the environmental and conservation movements, the use of these methods now usually falls under the purview of specially designated government agencies in most developed countries.

The 1970s saw a resurgence in the use of livestock guardian dogs and the development of new methods of predator control by sheep producers, many of them non-lethal. Donkeys and guard llamas have been used since the 1980s in sheep operations, using the same basic principle as livestock guardian dogs. Interspecific pasturing, usually with larger livestock such as cattle or horses, may help to deter predators, even if such species do not actively guard sheep. In addition to animal guardians, contemporary sheep operations may use non-lethal predator deterrents such as motion-activated lights and noisy alarms.

Economic Importance

Sheep are an important part of the global agricultural economy. However, their once vital status has been largely replaced by other livestock species, especially the pig, chicken, and cow. China, Australia, India, and Iran have the largest modern flocks, and serve both local and exportation needs for wool and mutton. Other countries such as New Zealand have smaller flocks but retain a large international economic impact due to their export of sheep products. Sheep also play a major role in many local economies, which may be niche markets focused on organic or sustainable agriculture and local food customers. Especially in developing countries, such flocks may be a part of subsistence agriculture rather than a system of trade. Sheep themselves may be a medium of trade in barter economies.

Wool supplied by Australian farmers to dealers (tonnes/quarter) has been in decline since 1990

Domestic sheep provide a wide array of raw materials. Wool was one of the first textiles, although in the late 20th century wool prices began to fall dramatically as the result of the popularity and cheap prices for synthetic fabrics. For many sheep owners, the cost of shearing is greater than the possible profit from the fleece, making subsisting on wool production alone practically impossible without farm subsidies. Fleeces are used as material in making alternative products such as wool insulation. In the 21st century, the sale of meat is the most profitable enterprise in the sheep industry, even though far less sheep meat is consumed than chicken, pork or beef.

Sheepskin is likewise used for making clothes, footwear, rugs, and other products. Byproducts from the slaughter of sheep are also of value: sheep tallow can be used in candle and soap making, sheep bone and cartilage has been used to furnish carved items such as dice and buttons as well as rendered glue and gelatin. Sheep intestine can be formed into sausage casings, and lamb intestine has been formed into surgical sutures, as well as strings for musical instruments and tennis rackets. Sheep droppings, which are high in cellulose, have even been sterilized and mixed with traditional pulp materials to make paper. Of all sheep byproducts, perhaps the most valuable is lanolin: the waterproof, fatty substance found naturally in sheep's wool and used as a base for innumerable cosmetics and other products.

Some farmers who keep sheep also make a profit from live sheep. Providing lambs for youth programs such as 4-H and competition at agricultural shows is often a dependable avenue for the sale of sheep. Farmers may also choose to focus on a particular breed of sheep in order to sell registered purebred animals, as well as provide a ram rental service for breeding. The most valuable sheep ever sold to date was a purebred Texel ram that fetched £231,000 at auction. The previous record holder was a Merino ram sold for £205,000 in 1989. A new option for deriving profit from live sheep is the rental of flocks for grazing; these "mowing services" are hired in order to keep unwanted vegetation down in public spaces and to lessen fire hazard.

Despite the falling demand and price for sheep products in many markets, sheep have distinct economic advantages when compared with other livestock. They do not require expensive housing, such as that used in the intensive farming of chickens or pigs. They are an efficient use of land; roughly six sheep can be kept on the amount that would suffice for a single cow or horse. Sheep can also consume plants, such as noxious weeds, that most other animals will not touch, and produce more young at a faster rate. Also, in contrast to most livestock species, the cost of raising sheep is not necessarily tied to the price of feed crops such as grain, soybeans and corn. Combined with the lower cost of quality sheep, all these factors combine to equal a lower overhead for sheep producers, thus entailing a higher profitability potential for the small farmer. Sheep are especially beneficial for independent producers, including family farms with limited resources, as the sheep industry is one of the few types of animal agriculture that has not been vertically integrated by agribusiness.

Food

Sheep meat and milk were one of the earliest staple proteins consumed by human civilization after the transition from hunting and gathering to agriculture. Sheep meat prepared for food is known as either mutton or lamb. "Mutton" is derived from the Old French *moton*, which was the word for sheep used by the Anglo-Norman rulers of much of the British Isles in the Middle Ages. This became the name for sheep meat in English, while the Old English word *sceap* was kept for the live animal. Throughout modern history, "mutton" has been limited to the meat of mature sheep usually at least two years of age; "lamb" is used for that of immature sheep less than a year.

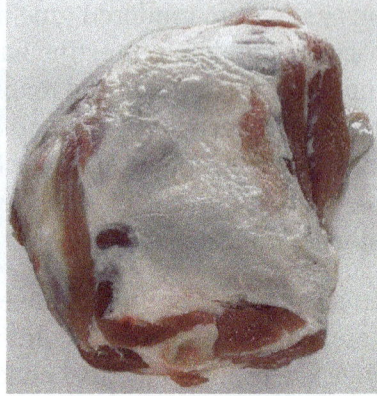

Shoulder of lamb

In the 21st century, the nations with the highest consumption of sheep meat are the Arab States of the Persian Gulf, New Zealand, Australia, Greece, Uruguay, the United Kingdom and Ireland. These countries eat 14–40 lbs (3–18 kg) of sheep meat per capita, per annum. Sheep meat is also popular in France, Africa (especially the Maghreb), the Caribbean, the rest of the Middle East, India, and parts of China. This often reflects a history of sheep production. In these countries in particular, dishes comprising alternative cuts and offal may be popular or traditional. Sheep testicles—called animelles or lamb fries—are considered a delicacy in many parts of the world. Perhaps the most unusual dish of sheep meat is the Scottish haggis, composed of various sheep innards cooked along with oatmeal and chopped onions inside its stomach. In comparison, countries such as the U.S. consume only a pound or less (under 0.5 kg), with Americans eating 50 pounds (22 kg) of pork and 65 pounds (29 kg) of beef. In addition, such countries rarely eat mutton, and may favor the more expensive cuts of lamb: mostly lamb chops and leg of lamb.

Though sheep's milk may be drunk rarely in fresh form, today it is used predominantly in cheese and yogurt making. Sheep have only two teats, and produce a far smaller volume of milk than cows. However, as sheep's milk contains far more fat, solids, and minerals than cow's milk, it is ideal for the cheese-making process. It also resists contamination during cooling better because of its much higher calcium content. Well-known cheeses made from sheep milk include the Feta of Bulgaria and Greece, Roquefort of France, Manchego from Spain, the Pecorino Romano (the Italian word for sheep is *pecore*) and Ricotta of Italy. Yogurts, especially some forms of strained yogurt, may also be made from sheep milk. Many of these products are now often made with cow's milk, especially when produced outside their country of origin. Sheep milk contains 4.8% lactose, which may affect those who are intolerant.

As with other domestic animals, the meat of uncastrated males is inferior in quality, especially as they grow. A "bucky" lamb is a lamb which was not castrated early enough, or which was castrated improperly (resulting in one testicle being retained). These lambs are worth less at market.

Science

Sheep are generally too large and reproduce too slowly to make ideal research subjects, and thus are not a common model organism. They have, however, played an influential role in some fields of science. In particular, the Roslin Institute of Edinburgh, Scotland used sheep for genetics research that produced groundbreaking results. In 1995, two ewes named Megan and Morag were the first mammals cloned from differentiated cells. A year later, a Finnish Dorset sheep named Dolly, dubbed "the world's most famous sheep" in *Scientific American*, was the first mammal to be cloned from an adult somatic cell. Following this, Polly and Molly were the first mammals to be simultaneously cloned and transgenic.

A cloned ewe named Dolly was a scientific landmark.

As of 2008, the sheep genome has not been fully sequenced, although a detailed genetic map has been published, and a draft version of the complete genome produced by assembling sheep DNA sequences using information given by the genomes of other mammals. In 2012, a transgenic sheep named "Peng Peng" was cloned by Chinese scientists, who spliced his genes with that of a roundworm (C. elegans) in order to increase production of fats healthier for human consumption.

In the study of natural selection, the population of Soay sheep that remain on the island of Hirta have been used to explore the relation of body size and coloration to reproductive success. Soay sheep come in several colors, and researchers investigated why the larger, darker sheep were in decline; this occurrence contradicted the rule of thumb that larger members of a population tend to be more successful reproductively. The feral Soays on Hirta are especially useful subjects because they are isolated.

Sheep are one of the few animals where the molecular basis of the diversity of male sexual preferences has been examined. However, this research has been controversial, and much publicity has been produced by a study at the Oregon Health and Science University that investigated the mechanisms that produce homosexuality in rams. Organizations such as PETA campaigned against the study, accusing scientists of trying to cure homosexuality in the sheep. OHSU and the involved scientists vehemently denied such accusations.

Domestic sheep are sometimes used in medical research, particularly for researching cardiovascular physiology, in areas such as hypertension and heart failure. Pregnant sheep are also a useful model for human pregnancy, and have been used to investigate the effects on fetal development of malnutrition and hypoxia. In behavioral sciences, sheep have been used in isolated cases for the study of facial recognition, as their mental process of recognition is qualitatively similar to humans.

Cultural Impact

Sheep have had a strong presence in many cultures, especially in areas where they form the most common type of livestock. In the English language, to call someone a sheep or ovine may allude that they are timid and easily led. In contradiction to this image, male sheep are often used as symbols of virility and power; the logos of the Los Angeles Rams football team and the Dodge Ram pickup truck allude to males of the bighorn sheep, *Ovis canadensis*.

The proverbial black sheep

A sheep on the Coat of arms of the Falkland Islands

Counting sheep is popularly said to be an aid to sleep, and some ancient systems of counting sheep persist today. Sheep also enter in colloquial sayings and idiom frequently with such phrases as "black sheep". To call an individual a black sheep implies that they are an odd or disreputable member of a group. This usage derives from the recessive trait that causes an occasional black lamb to be born into an entirely white flock. These black sheep were considered undesirable by shepherds, as black wool is not as commercially viable as white wool. Citizens who accept overbearing governments have been referred to by the Portmanteau neologism of sheeple. Somewhat differently, the adjective "sheepish" is also used to describe embarrassment.

Religion and Folklore

In antiquity, symbolism involving sheep cropped up in religions in the ancient Near East, the Mideast, and the Mediterranean area: Çatalhöyük, ancient Egyptian religion, the Cana'anite and Phoenician tradition, Judaism, Greek religion, and others. Religious symbolism and ritual involving sheep began with some of the first known faiths: Skulls of rams (along with bulls) occupied central placement in shrines at the Çatalhöyük settlement in 8,000 BCE. In Ancient Egyptian religion, the ram was the symbol of several gods: Khnum, Heryshaf and Amun (in his incarnation as a god of fertility). Other deities occasionally shown with ram features include the goddess Ishtar, the Phoenician god Baal-Hamon, and the Babylonian god Ea-Oannes. In Madagascar, sheep were not eaten as they were believed to be incarnations of the souls of ancestors.

Ancient Greek red-figure ram-head rhyton, ca. 340 BC

There are many ancient Greek references to sheep: that of Chrysomallos, the golden-fleeced ram, continuing to be told through into the modern era. Astrologically, Aries, the ram, is the first sign of the classical Greek zodiac, and the sheep is the eighth of the twelve animals associated with the 12-year cycle of in the Chinese zodiac, related to the Chinese calendar. In Mongolia, shagai are an ancient form of dice made from the cuboid bones of sheep that are often used for fortunetelling purposes.

Jesus is depicted as "The Good Shepherd", with the sheep being Christians

Sheep play an important role in all the Abrahamic faiths; Abraham, Isaac, Jacob, Moses, King David and the Islamic prophet Muhammad were all shepherds. According to the Biblical story of the Binding of Isaac, a ram is sacrificed as a substitute for Isaac after an angel stays Abraham's hand (in the Islamic tradition, Abraham was about to sacrifice Ishmael). Eid al-Adha is a major annual festival in Islam in which sheep (or other animals) are sacrificed in remembrance of this act. Sheep are occasionally sacrificed to commemorate important secular events in Islamic cultures. Greeks and Romans sacrificed sheep regularly in religious practice, and Judaism once sacrificed sheep as a Korban (sacrifice), such as the Passover lamb . Ovine symbols—such as the ceremonial blowing of a shofar—still find a presence in modern Judaic traditions. Followers of Christianity are collectively often referred to as a flock, with Christ as the Good Shepherd, and sheep are an element in the Christian iconography of the birth of Jesus. Some Christian saints are considered patrons of shepherds, and even of sheep themselves. Christ is also portrayed as the Sacrificial lamb of God (*Agnus Dei*) and Easter celebrations in Greece and Romania traditionally feature a meal of Paschal lamb. In many Christian traditions, a church leader is called the pastor, which is derived from the Latin word for shepherd.

Sheep are key symbols in fables and nursery rhymes like *The Wolf in Sheep's Clothing*, *Little Bo Peep*, *Baa, Baa, Black Sheep*, and *Mary Had a Little Lamb*; novels such as George Orwell's *Animal Farm* and Haruki Murakami's *A Wild Sheep Chase*; songs such as Bach's *Sheep may safely graze* (*Schafe können sicher weiden*) and Pink Floyd's *Sheep*, and poems like William Blake's "The Lamb".

Domestic Duck

Domesticated ducks are ducks that are raised for meat, eggs and down. Many ducks are also kept for show, as pets, or for their ornamental value. Almost all varieties of domesticated duck are descended from the mallard (*Anas platyrhynchos*), apart from the Muscovy duck (*Cairina moschata*).

Farming

A duck farm in Taiwan

Ducks have been farmed for thousands of years, possibly starting in Southeast Asia. In the Western world, they are not as popular as the chicken, because chickens have much more white lean meat and are easier to keep confined, making the total price much lower for chicken meat, whereas duck is comparatively expensive. While popular in *haute cuisine*, duck appears less frequently in the mass-market food industry and restaurants in the lower price range. However, ducks are more popular in China and there they are raised extensively.

Ducks are farmed for their meat, eggs, and down. A minority of ducks are also kept for *foie gras* production. In Vietnam, their blood is used in a food called *tiết canh*. Their eggs are blue-green to white, depending on the breed.

Ducks can be kept free range, in cages, in barns, or in batteries. Ducks enjoy access to swimming water, but do not require it to survive. They should be fed a grain and insect diet. It is a popular misconception that ducks should be fed bread; bread has limited nutritional value and can be deadly when fed to developing ducklings. Ducks should be monitored for avian influenza, as they are especially prone to infection with the dangerous H5N1 strain.

The females of many breeds of domestic ducks are unreliable at sitting their eggs and raising their young. Notable exceptions include the Rouen duck and especially the Muscovy duck. It has been a custom on farms for centuries to put duck eggs under broody hens for hatching; nowadays incubators are often used. However, young ducklings rely on their mothers for a supply of preen oil to make them waterproof; a chicken hen does not make as much preen oil as a female duck, and an incubator makes none. Once the duckling grows its own feathers, it will produce preen oil from the sebaceous gland near the base of its tail.

Pets and Ornamentals

Domestic duckling

Domesticated ducks can be kept as pets, in a garden or backyard, and with special accessories, have also been known to be kept in the house. They will often eat insects and slugs. A pond or deep water dish is recommended. If they are given access to a pond, they will dabble in the mud, dredging out and eating wildlife and frog spawn, and swallow adult frogs and toads up to the size of the British common frog *Rana temporaria*, as they have been bred to be much bigger than wild ducks, with a "hull length" (base of neck to base of tail) of up to 1 foot (30 cm) or more; the wild mallard's "hull length" is about 6 inches (15 cm). A coop should be provided for shelter from predators such as foxes, hawks, coyotes, and raccoons, as many breeds of domestic ducks cannot fly.

Ducks are also kept for their ornamental value. Breeds have been developed with crests and tufts or striking plumage. Exhibition shows are held in which ducks, along with other breeds of poultry, are exhibited in competition. These shows can be "open" (meaning any exhibitor who pays the required entry fee can enter), or "closed" (accessible only to members of a given group).

Cattle

Cattle—colloquially cows—are the most common type of large domesticated ungulates. They are a prominent modern member of the subfamily Bovinae, are the most widespread species of the genus *Bos*, and are most commonly classified collectively as *Bos taurus*. Cattle are raised as livestock for meat (beef and veal), as dairy animals for milk and other dairy products, and as draft animals (oxen or bullocks that pull carts, plows and other implements). Other products include leather and dung for manure or fuel. In some regions, such as parts of India, cattle have significant religious meaning. From as few as 80 progenitors domesticated in southeast Turkey about 10,500 years ago, according to an estimate from 2011, there are 1.4 billion cattle in the world. In 2009, cattle became one of the first livestock animals to have a fully mapped genome. Some consider cattle the oldest form of wealth, and cattle raiding consequently one of the earliest forms of theft.

Taxonomy

Cattle were originally identified as three separate species: *Bos taurus*, the European or "taurine" cattle (including similar types from Africa and Asia); *Bos indicus*, the zebu; and the extinct *Bos primigenius*, the aurochs. The aurochs is ancestral to both zebu and taurine cattle. Now, these have been reclassified as one species, *Bos taurus*, with three subspecies: *Bos taurus primigenius*, *Bos taurus indicus*, and *Bos taurus taurus*.

Żubroń, a cross between wisent and cattle

Complicating the matter is the ability of cattle to interbreed with other closely related species. Hybrid individuals and even breeds exist, not only between taurine cattle and zebu (such as the sanga cattle, *Bos taurus africanus*), but also between one or both of these and some other members of the genus *Bos* – yaks (the dzo or yattle), banteng, and gaur. Hybrids such as the beefalo breed can even occur between taurine cattle and either species of bison, leading some authors to consider them part of the genus *Bos*, as well. The hybrid origin of some types may not be obvious – for example, genetic testing of the Dwarf Lulu breed, the only taurine-type cattle in Nepal, found them to be a mix of taurine cattle, zebu, and yak. However, cattle cannot successfully be hybridized with more distantly related bovines such as water buffalo or African buffalo.

The aurochs originally ranged throughout Europe, North Africa, and much of Asia. In historical times, its range became restricted to Europe, and the last known individual died in Masovia, Poland, in about 1627. Breeders have attempted to recreate cattle of similar appearance to aurochs by crossing traditional types of domesticated cattle, creating the Heck cattle breed.

Etymology

Cattle did not originate as the term for bovine animals. It was borrowed from Anglo-Norman *catel*, itself from medieval Latin *capitale* 'principal sum of money, capital', itself derived in turn from Latin *caput* 'head'. *Cattle* originally meant movable personal property, especially livestock of any kind, as opposed to real property (the land, which also included wild or small free-roaming animals such as chickens — they were sold as part of the land). The word is a variant of *chattel* (a unit of personal property) and closely related to *capital* in the economic sense. The term replaced earlier Old English *feoh* 'cattle, property', which survives today as *fee* (cf. German: *Vieh*, Dutch: *vee*, Gothic: *faihu*).

The word "cow" came via Anglo-Saxon cū (plural cȳ), from Common Indo-European gwōus (genitive gwowés)="a bovine animal", compare Persian *gâv*, Sanskrit *go-*, Welsh *buwch*. The plural cȳ became *ki* or *kie* in Middle English, and an additional plural ending was often added, giving *kine*, *kien*, but also *kies*, *kuin* and others. This is the origin of the now archaic English plural, "kine". The Scots language singular is coo or cou, and the plural is "kye".

In older English sources such as the King James Version of the Bible, "cattle" refers to livestock, as opposed to "deer" which refers to wildlife. "Wild cattle" may refer to feral cattle or to undomesticated species of the genus *Bos*. Today, when used without any other qualifier, the modern meaning of "cattle" is usually restricted to domesticated bovines.

Terminology

In general, the same words are used in different parts of the world, but with minor differences in the definitions. The terminology described here contrasts the differences in definition between the United Kingdom and other British-influenced parts of world such as Canada, Australia, New Zealand, Ireland and the United States.

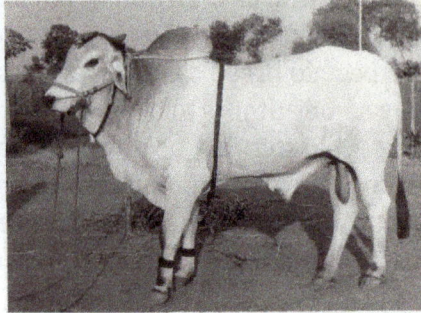
An Ongole bull

A Hereford bull

- An "intact" (i.e., not castrated) adult male is called a bull. A wild, young, un-marked bull is known as a "micky" in Australia. An unbranded bovine of either sex is called a "maverick" in the USA and Canada.

- An adult female that has had a calf (or two, depending on regional usage) is a cow.

- A young female before she has had a calf of her own and is under three years of age is called a heifer (/□h□fər/ *HEF*-ər). A young female that has had only one calf is occasionally called a first-calf heifer.

- Young cattle of both sexes are called calves until they are weaned, then weaners until they are a year old in some areas; in other areas, particularly with male beef cattle, they may be known as feeder calves or simply feeders. After that, they are referred to as yearlings or stirks if between one and two years of age.

- A castrated male is called a steer in the United States; older steers are often called bullocks in other parts of the world, but in North America this term refers to a young bull. Piker bullocks are micky bulls (uncastrated young male bulls) that were caught, castrated and then later lost. In Australia, the term "Japanese ox" is used for grain-fed steers in the weight range of 500 to 650 kg that are destined for the Japanese meat trade. In North America, draft cattle under four years old are called working steers. Improper or late castration on a bull results in it becoming a coarse steer known as a stag in Australia, Canada and New Zealand. In some countries, an incompletely castrated male is known also as a rig.

- A castrated male (occasionally a female or in some areas a bull) kept for draft purposes is called an ox (plural oxen); "ox" may also be used to refer to some carcass products from any adult cattle, such as ox-hide, ox-blood, oxtail, or ox-liver.

- A springer is a cow or heifer close to calving.

- In all cattle species, a female twin of a bull usually becomes an infertile partial intersex, and is called a freemartin.

- Neat (horned oxen, from which neatsfoot oil is derived), beef (young ox) and beefing (young animal fit for slaughtering) are obsolete terms, although poll, pollard or polled cattle are still terms in use for naturally hornless animals, or in some areas also for those that have been disbudded or dehorned.

- Cattle raised for human consumption are called beef cattle. Within the American beef cattle industry, the older term beef (plural beeves) is still used to refer to an animal of either sex. Some Australian, Canadian, New Zealand and British people use the term beast, especially for single animals when the sex is unknown.

- Cattle bred specifically for milk production are called milking or dairy cattle; a cow kept to provide milk for one family may be called a house cow or milker. A "fresh cow" is a dairy term for a cow or first-calf heifer who has recently given birth, or "freshened."

- The adjective applying to cattle in general is usually bovine. The terms "bull", "cow" and "calf" are also used by extension to denote the sex or age of other large animals, including whales, hippopotamuses, camels, elk and elephants.

Singular Terminology Issue

Cattle can only be used in the plural and not in the singular: it is a plurale tantum. Thus one may refer to "three cattle" or "some cattle", but not "one cattle". No universally used singular form in modern English of "cattle" exists, other than the sex- and age-specific terms such as cow, bull, steer and heifer. Historically, "ox" was not a sex-specific term for adult cattle, but generally this is now used only for draft cattle, especially adult castrated males. The term is also incorporated into the names of other species, such as the musk ox and "grunting ox" (yak), and is used in some areas to describe certain cattle products such as ox-hide and oxtail.

A Brahman calf

"Cow" is in general use as a singular for the collective "cattle", despite the objections by those who insist it to be a female-specific term. Although the phrase "that cow is a bull" is absurd from a lexicographic standpoint, the word "cow" is easy to use when a singular is needed and the sex is unknown or irrelevant – when "there is a cow in the road", for example. Further, any herd of fully mature cattle in or near a pasture is statistically likely to consist mostly of cows, so the term is probably accurate even in the restrictive sense. Other than the few bulls needed for breeding, the vast majority of male cattle are castrated as calves and slaughtered for meat before the age of three years. Thus, in a pastured herd, any calves or herd bulls usually are clearly distinguishable from the cows due to distinctively different sizes and clear anatomical differences. Merriam-Webster, a US dictionary, recognizes the sex-nonspecific use of "cow" as an alternate definition, whereas Collins, a UK dictionary, does not.

Colloquially, more general nonspecific terms may denote cattle when a singular form is needed. Australian, New Zealand and British farmers use the term "beast" or "cattle beast". "Bovine" is also used in Britain. The term "critter" is common in the western United States and Canada, particularly when referring to young cattle. In some areas of the American South (particularly the Appalachian region), where both dairy and beef cattle are present, an individual animal was once called a "beef critter", though that term is becoming archaic.

Other Terminology

Cattle raised for human consumption are called "beef cattle". Within the beef cattle industry in parts of the United States, the term "beef" (plural "beeves") is still used in its archaic sense to refer to an animal of either sex. Cows of certain breeds that are kept for the milk they give are called "dairy cows" or "milking cows" (formerly "milch cows"). Most young male offspring of dairy cows are sold for veal, and may be referred to as veal calves.

The term "dogies" is used to describe orphaned calves in the context of ranch work in the American West, as in "Keep them dogies moving". In some places, a cow kept to provide milk for one family is called a "house cow". Other obsolete terms for cattle include "neat" (this use survives in "neatsfoot oil", extracted from the feet and legs of cattle), and "beefing" (young animal fit for slaughter).

An onomatopoeic term for one of the most common sounds made by cattle is "moo" (also called *lowing*). There are a number of other sounds made by cattle, including calves *bawling*, and bulls *bellowing*. Bawling is most common for cows after weaning of a calf. The bullroarer makes a sound similar to a bull's territorial call.

Characteristics

Anatomy

Cattle are large quadrupedal ungulate mammals with cloven hooves. Most breeds have horns, which can be as large as the Texas Longhorn or small like a scur. Careful genetic selection has allowed polled (hornless) cattle to become widespread.

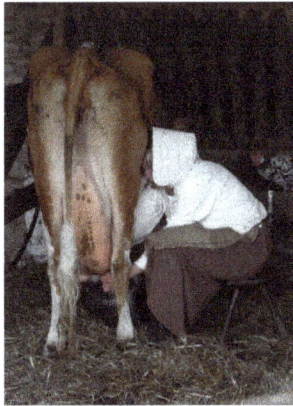

Dairy farming and the milking of cattle was once performed largely by hand, but is now usually replaced by machine

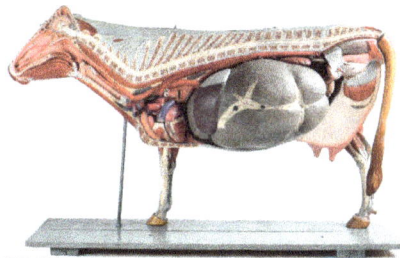

Anatomy model of a bovine (cow)

Cattle are ruminants, meaning their digestive system is highly specialized to allow the use of poorly digestible plants as food. Cattle have one stomach with four compartments, the rumen, reticulum, omasum, and abomasum, with the rumen being the largest compartment. The reticulum, the smallest compartment, is known as the "honeycomb". Cattle sometimes consume metal objects which are deposited in the reticulum and irritation from the metal objects causes hardware disease. The omasum's main function is to absorb water and nutrients from the digestible feed. The omasum is known as the "many plies". The abomasum is like the human stomach; this is why it is known as the "true stomach".

Cattle are known for regurgitating and re-chewing their food, known as cud chewing, like most ruminants. While the animal is feeding, the food is swallowed without being chewed and goes into the rumen for storage until the animal can find a quiet place to continue the digestion process. The food is regurgitated, a mouthful at a time, back up to the mouth, where the food, now called the cud, is chewed by the molars, grinding down the course vegetation to small particles. The cud is then swallowed again and further digested by specialized microorganisms in the rumen. These microbes are primarily responsible for decomposing cellulose and other carbohydrates into volatile fatty acids cattle use as their primary metabolic fuel. The microbes inside the rumen also synthesize amino acids from non-protein nitrogenous sources, such as urea and ammonia. As these microbes reproduce in the rumen, older generations die and their cells continue on through the digestive tract. These cells are then partially digested in the small intestines, allowing cattle to gain a high-quality protein source. These features allow cattle to thrive on grasses and other tough vegetation.

Gestation and Size

The gestation period for a cow is about nine months long. A newborn calf's size can vary among breeds, but a typical calf weighs between 25 to 45 kg (55 to 99 lb). Adult size and weight vary significantly among breeds and sex. The world record for the heaviest bull was 1,740 kg (3,840 lb), a Chianina named Donetto, when he was exhibited at the Arezzo show in 1955. The heaviest steer was eight-year-old 'Old Ben', a Shorthorn/Hereford cross weighing in at 2,140 kg (4,720 lb) in 1910. Steers are generally killed before reaching 750 kg (1,650 lb). Breeding stock may be allowed a longer lifespan, occasionally living as long as 25 years. The oldest recorded cow, Big Bertha, died at the age of 48 in 1993.

Udder

A cow's udder contains two pairs of mammary glands, (commonly referred to as *teats*) creating four "quarters". The front ones are referred to as *fore quarters* and the rear ones *rear quarters*.

Male Genitalia

Bulls become fertile at about seven months of age. Their fertility is closely related to the size of their testicles, and one simple test of fertility is to measure the circumference of the scrotum: a young bull is likely to be fertile once this reaches 28 centimetres (11 in); that of a fully adult bull may be over 40 centimetres (16 in).

Bulls have a fibro-elastic penis. Given the small amount of erectile tissue, there is little enlargement after erection. The penis is quite rigid when non-erect, and becomes even more rigid during erection. Protrusion is not affected much by erection, but more by relaxation of the retractor penis muscle and straightening of the sigmoid flexure.

Weight

The weight of adult cattle always depends on the breed. Smaller kinds, such as Dexter and Jersey adults, range between 272 to 454 kg (600 to 1,000 lb). Large Continental breeds, such as Charolais, Marchigiana, Belgian Blue and Chianina, adults range from 635 to 1,134 kg (1,400 to 2,500 lb). British breeds, such as Hereford, Angus, and Short-horn, mature between 454 to 907 kg (1,000 to 2,000 lb), occasionally higher, particularly with Angus and Hereford.

Bulls will be a bit larger than cows of the same breed by a few hundred kilograms. Chianina bulls can weigh up to 1,500 kg (3,300 lb); British bulls, such as Angus and Hereford, can weigh as little as 907 kg (2,000 lb) to as much as 1,361 kg (3,000 lb).

It is difficult to generalize or average out the weight of all cattle because different kinds have different averages of weights. However, according to some sources, the average weight of all cattle is 753 kg (1,660 lb). Finishing steers in the feedlot average about 640 kg (1,410 lb); cows about 725 kg (1,600 lb), and bulls about 1,090 kg (2,400 lb).

In the United States, the average weight of beef cattle has steadily increased, especially since the 1970s, requiring the building of new slaughterhouses able to handle larger carcasses. New packing plants in the 1980s stimulated a large increase in cattle weights. Before 1790 beef cattle averaged only 160 kg (350 lb) net; and thereafter weights climbed steadily.

Cognition

In laboratory studies, young cattle are able to memorize the locations of several food sources and retain this memory for at least 8 hours, although this declined after 12 hours. Fifteen-month-old heifers learn more quickly than adult cows which have had either one or two calvings, but their longer-term memory is less stable. Mature cattle perform well in spatial learning tasks and have a good long-term memory in these tests. Cattle tested in a radial arm maze are able to remember the locations of high-quality food for at least 30 days. Although they initially learn to avoid low-quality food, this memory diminishes over the same duration. Under less artificial testing conditions, young cattle showed they were able to remember the location of feed for at least 48 days. Cattle can make an association between a visual stimulus and food within 1 day – memory of this association can be retained for 1 year, despite a slight decay.

Calves are capable of discrimination learning and adult cattle compare favourably with small mammals in their learning ability in the Closed-field Test.

They are also able to discriminate between familiar individuals, and among humans. Cattle can tell the difference between familiar and unfamiliar animals of the same species (conspecifics). Studies show they behave less aggressively toward familiar individuals when they are forming a new group. Calves can also discriminate between humans based on previous experience, as shown by approaching those who handled them positively and avoiding those who handled them aversively. Although cattle can discriminate between humans by their faces alone, they also use other cues such as the color of clothes when these are available.

In audio play-back studies, calves prefer their own mother's vocalizations compared to the vocalizations of an unfamiliar mother.

In laboratory studies using images, cattle can discriminate between images of the heads of cattle and other animal species. They are also able to distinguish between familiar and unfamiliar conspecifics. Furthermore, they are able to categorize images as familiar and unfamiliar individuals.

When mixed with other individuals, cloned calves from the same donor form subgroups, indicating that kin discrimination occurs and may be a basis of grouping behaviour. It has also been shown using images of cattle that both artificially inseminated and cloned calves have similar cognitive capacities of kin and non-kin discrimination.

Cattle can recognize familiar individuals. Visual individual recognition is a more complex mental process than visual discrimination. It requires the recollection of the learned idiosyncratic identity of an individual that has been previously encountered and the formation of a mental representation. By using 2-dimensional images of the heads of one cow (face, profiles, ¾ views), all the tested heifers showed individual recognition of familiar and unfamiliar individuals from their own breed. Furthermore, almost all the heifers recognized unknown individuals from different breeds, although this was achieved with greater difficulty. Individual recognition was most difficult when the visual features of the breed being tested were quite different from the breed in the image, for example, the breed being tested had no spots whereas the image was of a spotted breed.

Cattle use visual/brain lateralisation in their visual scanning of novel and familiar stimuli. Domestic cattle prefer to view novel stimuli with the left eye, i.e. using the right brain hemisphere (similar to horses, Australian magpies, chicks, toads and fish) but use the right eye, i.e. using the left hemisphere, for viewing familiar stimuli.

Temperament and Emotions

In cattle, temperament can affect production traits such as carcass and meat quality or milk yield as well as affecting the animal's overall health and reproduction. Cattle temperament is defined as "the consistent behavioral and physiological difference observed between individuals in response to a stressor or environmental challenge and is used to describe the relatively stable difference in the behavioral predisposition of an animal, which can be related to psychobiological mechanisms". Generally, cattle temperament is assumed to be multidimensional. Five underlying categories of temperament traits have been proposed:

- shyness-boldness

- exploration-avoidance

- activity

- aggressiveness

- sociability

In a study on Holstein–Friesian heifers learning to press a panel to open a gate for access to a food reward, the researchers also recorded the heart rate and behavior of the heifers when moving along the race towards the food. When the heifers made clear improvements in learning, they had higher heart rates and tended to move more vigorously along the race. The researchers concluded this was an indication that cattle may react emotionally to their own learning improvement.

Negative emotional states are associated with a bias toward negative (pessimistic) responses towards ambiguous cues in judgement tasks – as encapsulated in the question of "is the glass half empty or half full?". After separation from their mothers, Holstein calves showed such a cognitive bias indicative of low mood. A similar study showed that after hot-iron disbudding (dehorning), calves had a similar negative bias indicating that post-operative pain following this routine procedure results in a negative change in emotional state.

In studies of visual discrimination, the position of the ears has been used as an indicator of emotional state. When cattle are stressed, this can be recognised by other cattle as it is communicated by alarm substances in the urine.

Cattle are very gregarious and even short-term isolation is considered to cause severe psychological stress. When Aubrac and Fresian heifers are isolated, they increase their vocalizations and experience increased heart rate and plasma cortisol concentrations. These physiological changes are greater in Aubracs. When visual contact is re-instated, vocalisations rapidly decline, regardless of the familiarity of the returning cattle, however, heart rate decreases are greater if the returning cattle are familiar to the previously-isolated individual. Mirrors have been used to reduce stress in isolated cattle.

Senses

Cattle use all of the five widely recognized sensory modalities. These can assist in some complex behavioural patterns, for example, in grazing behaviour. Cattle eat mixed diets, but when given the opportunity, show a partial preference of approximately 70% clover and 30% grass. This preference has a diurnal pattern, with a stronger preference for clover in the morning, and the proportion of grass increasing towards the evening.

Vision

Vision is the dominant sense in cattle and they obtain almost 50% of their information visually.

Cattle are a prey animal and to assist predator detection, their eyes are located on the sides of their head rather than the front. This gives them a wide field of view of 330° but limits binocular vision (and therefore stereopsis) to 30° to 50° compared to 140° in humans. This means they have a blind spot directly behind them. Cattle have good visual acuity (1/20) but compared to humans, the visual accommodation of cattle is poor.

Cattle have two kinds of color receptors in the cone cells of their retinas. This means that cattle are dichromatic, as are most other non-primate land mammals. There are two to three rods per cone in the fovea centralis but five to six near the optic papilla. Cattle can distinguish long wavelength colors (yellow, orange and red) much better than the shorter wavelengths (blue, grey and green). Calves are able to discriminate between long (red) and short (blue) or medium (green) wavelengths, but have limited ability to discriminate between the short and medium. They also approach handlers more quickly under red light. Whilst having good color sensitivity, it is not as good as humans or sheep.

A common misconception about cattle (particularly bulls) is that they are enraged by the color red (something provocative is often said to be "like a red flag to a bull"). This is a myth. In bullfighting, it is the movement of the red flag or cape that irritates the bull and incites it to charge.

Taste

Cattle have a well-developed sense of taste and can distinguish the four primary tastes (sweet, salty, bitter and sour). They possess around 20,000 taste buds. The strength of taste perception depends on the individual's current food requirements. They avoid bitter-tasting foods (potentially toxic) and have a marked preference for sweet (high calorific value) and salty foods (electrolyte balance). Their sensitivity to sour-tasting foods helps them to maintain optimal ruminal pH.

Plants have low levels of sodium and cattle have developed the capacity of seeking salt by taste and smell. If cattle become depleted of sodium salts, they show increased loco-motion directed to searching for these. To assist in their search, the olfactory and gustatory receptors able to detect minute amounts of sodium salts increase their sensitivity as biochemical disruption develops with sodium salt depletion.

Audition

Cattle hearing ranges from 23 Hz to 35 kHz. Their frequency of best sensitivity is 8 kHz and they have a lowest threshold of −21 db (re 20 μN/m^{-2}), which means their hearing is more acute than horses (lowest threshold of 7 db). Sound localization acuity thresholds are an average of 30°. This means that cattle are less able to localise sounds compared to goats (18°), dogs (8°) and humans (0.8°). Because cattle have a broad foveal fields of view covering almost the entire horizon, they may not need very accurate locus information from their auditory systems to direct their gaze to a sound source.

Vocalisations are an important mode of communication amongst cattle and can provide information on the age, sex, dominance status and reproductive status of the caller. Calves can recognize their mothers using vocal and vocal behaviour may play a role by indicating estrus and competitive display by bulls.

Olfaction and Gustation

Cattle have a range of odiferous glands over their body including interdigital, infraorbital, inguinal and sebaceous glands, indicating that olfaction probably plays a large role in their social life. Both the primary olfactory system using the olfactory bulbs, and the secondary olfactory system using the vomeronasal organ are used. This latter olfactory system is used in the flehmen response. There is evidence that when cattle are stressed, this can be recognised by other cattle and this is communicated by alarm substances in the urine. The odour of dog faeces induces behavioural changes prior to cattle feeding, whereas the odours of urine from either stressed or non-stressed conspecifics and blood have no effect.

Several senses are used in social relationships between cattle

In the laboratory, cattle can be trained to recognise conspecific individuals using olfaction only.

In general, cattle use their sense of smell to "expand" on information detected by other sensory modalities. However, in the case of social and reproductive behaviours, olfaction is a key source of information.

Touch

Cattle have tactile sensations detected mainly by mechanoreceptors, thermoreceptors and nociceptors in the skin and muzzle. These are used most frequently when cattle explore their environment.

Magnetoreception

There is conflicting evidence for magnetoreception in cattle. One study reported that resting and grazing cattle tend to align their body axes in the geomagnetic North-South (N-S) direction. In a follow-up study, cattle exposed to various magnetic fields directly beneath or in the vicinity of power lines trending in various magnetic directions exhibited distinct patterns of alignment. However, in 2011, a group of Czech researchers reported their failed attempt to replicate the finding using Google Earth images.

Behaviour

Under natural conditions, calves stay with their mother until weaning at 8 to 11 months. Heifer and bull calves are equally attached to their mothers in the first few months of life. Cattle are considered to be "hider" type animals, but in the artificial environment of small calving pens, close proximity between cow and calf is maintained by the mother at the first three calvings but this changes to being mediated by the calf after these. Primiparous dams show a higher incidence of abnormal maternal behaviour.

Video of a calf suckling

Beef-calves reared on the range suckle an average of 5.0 times every 24 hours with an average total time of 46 min spent suckling. There is a diurnal rhythm in suckling activity with peaks between 05:00–07:00, 10:00–13:00 and 17:00–21:00.

Studies on the natural weaning of zebu cattle (*Bos indicus*) have shown that the cow weans her calves over a 2-week period, but after that, she continues to show strong affiliatory behaviour with her offspring and preferentially chooses them for grooming and as grazing partners for at least 4–5 years.

Reproductive Behaviour

Semi-wild Highland cattle heifers first give birth at 2 or 3 years of age and the timing of birth is synchronized with increases in natural food quality. Average calving interval is 391 days, and calving mortality within the first year of life is 5%.

Dominance and Leadership

One study showed that over a 4-year period, dominance relationships within a herd of semi-wild highland cattle were very firm. There were few overt aggressive conflicts and the majority of disputes were settled by agonistic (non-aggressive, competitive) behaviours that involved no physical contact between opponents (e.g. threatening and spontaneous withdrawing). Such agonistic behaviour reduces the risk of injury. Dominance status depended on age and sex, with older animals generally being dominant to young ones and males dominant to females. Young bulls gained superior dominance status over adult cows when they reached about 2 years of age.

As with many animal dominance hierarchies, dominance-associated aggressiveness does not correlate with rank position, but is closely related to rank distance between individuals.

Dominance is maintained in several ways. Cattle often engage in mock fights where they test each other's strength in a non-aggressive way. Licking is primarily performed by subordinates and received by dominant animals. Mounting is a playful behaviour shown by calves of both sexes and by bulls but not by cows, however, this is not a dominance related behaviour as has been found in other species.

The horns of cattle are "honest signals" used in mate selection. Furthermore, horned cattle attempt to keep greater distances between themselves and have fewer physical interactions than hornless cattle. This leads to more stable social relationships.

In calves, the frequency of agonistic behavior decreases as space allowance increases, but this does not occur for changes in group size. However, in adult cattle, the number of agonistic encounters increases as the group size increases.

Grazing Behaviour

When grazing, cattle vary several aspects of their bite, i.e. tongue and jaw movements, depending on characteristics of the plant they are eating. Bite area decreases with the density of the plants but increases with their height. Bite area is determined by the sweep of the tongue; in one study observing 750-kilogram (1,650 lb) steers, bite area reached a maximum of approximately 170 cm^2 (30 sq in). Bite depth increases with the height of the plants. By adjusting their behaviour, cattle obtain heavier bites in swards that are tall and sparse compared with short, dense swards of equal mass/area. Cattle adjust other aspects of their grazing behaviour in relation to the available food; foraging velocity decreases and intake rate increases in areas of abundant palatable forage.

Cattle avoid grazing areas contaminated by the faeces of other cattle more strongly than they avoid areas contaminated by sheep, but they do not avoid pasture contaminated by rabbit faeces.

Genetics

In the 24 April 2009, edition of the journal *Science*, a team of researchers led by the National Institutes of Health and the US Department of Agriculture reported having mapped the bovine genome. The scientists found cattle have about 22,000 genes, and 80% of their genes are shared with humans, and they share about 1000 genes with dogs and rodents, but are not found in humans. Using this bovine "HapMap", researchers can track the differences between the breeds that affect the quality of meat and milk yields.

Behavioral traits of cattle can be as heritable as some production traits, and often, the two can be related. The heritability of fear varies markedly in cattle from low (0.1) to high (0.53); such high variation is also found in pigs and sheep, probably due to differences in the methods used. The heritability of temperament (response to isolation during handling) has been calculated as 0.36 and 0.46 for habituation to handling. Rangeland assessments show that the heritability of aggressiveness in cattle is around 0.36.

Quantitative trait loci (QTLs) have been found for a range of production and behavioral characteristics for both dairy and beef cattle.

Domestication and Husbandry

Cattle occupy a unique role in human history, having been domesticated since at least the early neolithic age.

Texas Longhorns are a US breed

Archeozoological and genetic data indicate that cattle were first domesticated from wild aurochs (*Bos primigenius*) approximately 10,500 years ago. There were two major areas of domestication: one in the area that is now Turkey, giving rise to the taurine line, and a second in the area that is now Pakistan, resulting in the indicine line. Modern mitochondrial DNA variation indicates the taurine line may have arisen from as few as 80 aurochs tamed in the upper reaches of Mesopotamia near the villages of Çayönü Tepesi in southeastern Turkey and Dja'de el-Mughara in northern Iraq.

Although European cattle are largely descended from the taurine lineage, gene flow from African cattle (partially of indicine origin) contributed substantial genomic components to both southern European cattle breeds and their New World descendants. A study on 134 breeds showed that modern taurine cattle originated from Africa, Asia, North and South America, Australia, and Europe. Some researchers have suggested that African taurine cattle are derived from a third independent domestication from North African aurochsen.

Usage as Money

As early as 9000 BC both grain and cattle were used as money or as *barter* (Davies) (the *first grain remains* found, considered to be evidence of pre-agricultural practice date to 17,000 BC). Some evidence also exists to suggest that other animals, such as camels and goats, may have been used as currency in some parts of the world. One of the advantages of using cattle as currency is that it allows the seller to set a fixed price. It even created the standard pricing. For example, two chickens were traded for one cow as cows were deemed to be more valuable than chickens.

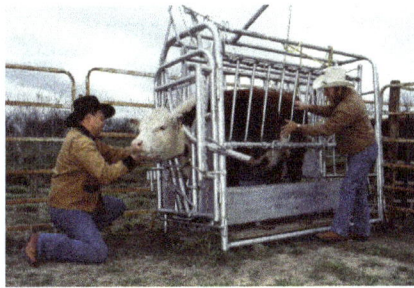

This Hereford is being inspected for ticks; cattle are often restrained or confined in cattle crushes (squeeze chutes) when given medical attention.

Modern Husbandry

Cattle are often raised by allowing herds to graze on the grasses of large tracts of rangeland. Raising cattle in this manner allows the use of land that might be unsuitable for growing crops. The most common interactions with cattle involve daily feeding, cleaning and milking. Many routine husbandry practices involve ear tagging, dehorning, loading, medical operations, vaccinations and hoof care, as well as training for agricultural shows and preparations. Also, some cultural differences occur in working with cattle; the cattle husbandry of Fulani men rests on behavioural techniques, whereas in Europe, cattle are controlled primarily by physical means, such as fences. Breeders use cattle husbandry to reduce *M. bovis* infection susceptibility by selective breeding and maintaining herd health to avoid concurrent disease.

This young bovine has a nose ring to prevent it from suckling, which is usually to assist in weaning.

Cattle are farmed for beef, veal, dairy, and leather, and they are less commonly used for conservation grazing, simply to maintain grassland for wildlife – for example, in Epping Forest, England. They are often used in some of the most wild places for livestock. Depending on the breed, cattle can survive on hill grazing, heaths, marshes, moors and semidesert. Modern cattle are more commercial than older breeds and, having become more specialized, are less versatile. For this reason, many smaller farmers still favor old breeds, such as the Jersey dairy breed. In Portugal, Spain, southern France and some Latin American countries, bulls are used in the activity of bullfighting; *Jallikattu* in India is a bull taming sport radically different from European bullfighting, humans are unarmed and bulls are not killed. In many other countries bullfighting is illegal. Other activities such as bull riding are seen as part of a rodeo, especially in North America. Bull-leaping, a central ritual in Bronze Age Minoan culture, still exists in south-western France. In modern times, cattle are also entered into agricultural competitions. These competitions can involve live cattle or cattle carcases in hoof and hook events.

In terms of food intake by humans, consumption of cattle is less efficient than of grain or vegetables with regard to land use, and hence cattle grazing consumes more area than such other agricultural production when raised on grains. Nonetheless, cattle and other forms of domesticated animals can sometimes help to use plant resources in areas not easily amenable to other forms of agriculture.

Sleep

The average sleep time of a domestic cow is about four hours a day. Cattle do have a stay apparatus, but do not sleep standing up, they lie down to sleep deeply. In spite of the urban legend, cows cannot be tipped over by people pushing on them.

Economy

The meat of adult cattle is known as beef, and that of calves is veal. Other animal parts are also used as food products, including blood, liver, kidney, heart and oxtail. Cattle also produce milk, and dairy cattle are specifically bred to produce the large quantities of milk processed and sold for human consumption. Cattle today are the basis of a multibillion-dollar industry worldwide. The international trade in beef for 2000 was over $30 billion and represented only 23% of world beef production. The production of milk, which is also made into cheese, butter, yogurt, and other dairy products, is comparable in economic size to beef production, and provides an important part of the food supply for many of the world's people. Cattle hides, used for leather to make shoes, couches and clothing, are another widespread product. Cattle remain broadly used as draft animals in many developing countries, such as India. Cattle are also used in some sporting games, including rodeo and bullfighting.

Holstein cattle are the primary dairy breed, bred for high milk production.

Cattle Meat Production

Cattle meat production (kt)				
	2008	**2009**	**2010**	**2011**
Argentina	3132	3378	2630	2497
Australia	2132	2124	2630	2420
Brazil	9024	9395	9115	9030
China	5841	6060	6244	6182
Germany	1199	1190	1205	1170
Japan	520	517	515	500
USA	12163	11891	12046	11988

Source: Helgi Library, World Bank, FAOSTAT

About half the world's meat comes from cattle.

Dairy

Certain breeds of cattle, such as the Holstein-Friesian, are used to produce milk, which can be processed into dairy products such as milk, cheese or yogurt. Dairy cattle are usually kept on specialized dairy farms designed for milk production. Most cows are milked twice per day, with milk processed at a dairy, which may be onsite at the farm or the milk may be shipped to a dairy plant for eventual sale of a dairy product. For dairy cattle to continue producing milk, they must give birth to one calf per year. If the calf is male, it generally is slaughtered at a young age to produce veal. They will continue to produce milk until three weeks before birth. Over the last fifty years, dairy farming has become more intensive to increase the yield of milk produced by each cow. The Holstein-Friesian is the breed of dairy cow most common in the UK, Europe and the United States. It has been bred selectively to produce the highest yields of milk of any cow. Around 22 litres per day is average in the UK.

Hides

Most cattle are not kept solely for hides, which are usually a by-product of beef production. Hides are most commonly used for leather which can be made into a variety of product including shoes. In 2012 India was the world's largest producer of cattle hides.

Feral Cattle

Feral cattle are defined as being 'cattle that are not domesticated or cultivated'. Populations of feral cattle are known to come from and exist in: Australia, United States of America, Colombia, Argentina, Spain, France and many islands, including New Guinea, Hawaii, Galapagos, Juan Fernández Islands, Hispaniola (Dominican Republic and Haiti), Tristan da Cunha and Île Amsterdam, two islands of Kuchinoshima and Kazura Island next to Naru Island in Japan. Chillingham cattle is sometimes regarded as a feral breed. Aleutian wild cattles can be found on Aleutian Islands. The "Kinmen cattle" which is dominantly found on Kinmen Island, Taiwan is mostly domesticated while smaller portion of the population is believed to live in the wild due to accidental releases.

Environmental Impact

Cattle near the Bruneau River in Elko County, Nevada

Cattle in dry landscape north of Alice Springs, Australia (CSIRO)

A report from the Food and Agriculture Organization (FAO) states that the livestock sector is "responsible for 18% of greenhouse gas emissions". The report concludes, unless changes are made, the damage thought to be linked to livestock may more than double by 2050, as demand for meat increases. Another concern is manure, which if not well-managed, can lead to adverse environmental consequences. However, manure also is a valuable source of nutrients and organic matter when used as a fertilizer. Manure was used as a fertilizer on about 15.8 million acres of US cropland in 2006, with manure from cattle accounting for nearly 70% of manure applications to soybeans and about 80% or more of manure applications to corn, wheat, barley, oats and sorghum. Substitution of manure for synthetic fertilizers in crop production can be environmentally significant, as between 43 and 88 megajoules of fossil fuel energy would be used per kg of nitrogen in manufacture of synthetic nitrogenous fertilizers.

One of the cited changes suggested to reduce greenhouse gas emissions is intensification of the livestock industry, since intensification leads to less land for a given level of production. This assertion is supported by studies of the US beef production system, suggesting practices prevailing in 2007 involved 8.6% less fossil fuel use, 16.3% less greenhouse gas emissions, 12.1% less water use, and 33.0% less land use, per unit mass of beef produced, than those used in 1977. The analysis took into account not only practices in feedlots, but also feed production (with less feed needed in more intensive production systems), forage-based cow-calf operations and backgrounding before cattle enter a feedlot (with more beef produced per head of cattle from those sources, in more intensive systems), and beef from animals derived from the dairy industry.

The number of American cattle kept in confined feedlot conditions fluctuates. From 1 January 2002 through 1 January 2012, there was no significant overall upward or downward trend in the number of US cattle on feed for slaughter, which averaged about 14.046 million head over that period. Previously, the number had increased; it was 12.453 million in 1985. Cattle on feed (for slaughter) numbered about 14.121 million on 1 January 2012, i.e. about 15.5% of the estimated inventory of 90.8 million US cattle (including calves) on that date. Of the 14.121 million, US cattle on feed (for slaughter) in operations with 1000 head or more were estimated to number 11.9 million. Cattle feedlots in this size category correspond to the regulatory definition of "large" concentrated animal feeding operations (CAFOs) for cattle other than mature dairy cows or veal calves. Significant numbers of dairy, as well as beef cattle, are confined in CAFOs, defined as "new and existing operations which stable or confine and feed or maintain for a total of 45 days or more in any 12-month period more than the number of animals specified" where "[c]rops, vegetation, forage growth, or post-harvest residues are not sustained in the normal growing season over any portion of the lot or facility." They may be designated as small, medium and large. Such designation of cattle CAFOs is according to cattle type (mature dairy cows, veal calves or other) and cattle numbers, but medium CAFOs are so designated only if they meet certain discharge criteria, and small CAFOs are designated only on a case-by-case basis.

A CAFO that discharges pollutants is required to obtain a permit, which requires a plan to manage nutrient runoff, manure, chemicals, contaminants, and other wastewater pursuant to the US Clean Water Act. The regulations involving CAFO permitting have been extensively litigated. Commonly, CAFO wastewater and manure nutrients

are applied to land at agronomic rates for use by forages or crops, and it is often assumed that various constituents of wastewater and manure, e.g. organic contaminants
and pathogens, will be retained, inactivated or degraded on the land with application
at such rates; however, additional evidence is needed to test reliability of such assump-
tions . Concerns raised by opponents of CAFOs have included risks of contaminated
water due to feedlot runoff, soil erosion, human and animal exposure to toxic chemicals, development of antibiotic resistant bacteria and an increase in *E. coli* contamination. While research suggests some of these impacts can be mitigated by developing
wastewater treatment systems and planting cover crops in larger setback zones, the
Union of Concerned Scientists released a report in 2008 concluding that CAFOs are
generally unsustainable and externalize costs.

An estimated 935,000 cattle operations were operating in the USA in 2010. In 2001,
the US Environmental Protection Agency (EPA) tallied 5,990 cattle CAFOs then regulated, consisting of beef (2,200), dairy (3,150), heifer (620) and veal operations (20).
Since that time, the EPA has established CAFOs as an enforcement priority. EPA enforcement highlights for fiscal year 2010 indicated enforcement actions against 12 cattle CAFOs for violations that included failures to obtain a permit, failures to meet the
terms of a permit, and discharges of contaminated water.

Cattle grazing in a high-elevation environment at the Big Pasture Plateau, Slovenia

Grazing by cattle at low intensities can create a favourable environment for native
herbs and forbs; in many world regions, though, cattle are reducing biodiversity due to
overgrazing. A survey of refuge managers on 123 National Wildlife Refuges in the US
tallied 86 species of wildlife considered positively affected and 82 considered negatively affected by refuge cattle grazing or haying. Proper management of pastures, notably
managed intensive rotational grazing and grazing at low intensities can lead to less use
of fossil fuel energy, increased recapture of carbon dioxide, fewer ammonia emissions
into the atmosphere, reduced soil erosion, better air quality, and less water pollution.

Some microbes in the cattle gut carry out anaerobic process known as methanogenesis,
which produces methane. Cattle and other livestock emit about 80 to 93 Tg of methane per year, accounting for an estimated 37% of anthropogenic methane emissions,
and additional methane is produced by anaerobic fermentation of manure in manure
lagoons and other manure storage structures. The 100-year global warming potential

of methane, including effects on ozone and stratospheric water vapor, is 25 times as great as that of carbon dioxide. Methane's effect on global warming is correlated with changes in atmospheric methane content, not with emissions. The net change in atmospheric methane content was recently about 1 Tg per year, and in some recent years there has been no increase in atmospheric methane content. Mitigation options for reducing methane emission from ruminant enteric fermentation include genetic selection, immunization, rumen defaunation, diet modification and grazing management, among others. While cattle fed forage actually produce more methane than grain-fed cattle, the increase may be offset by the increased carbon recapture of pastures, which recapture three times the CO_2 of cropland used for grain.

Health

The veterinary discipline dealing with cattle and cattle diseases (bovine veterinary) is called buiatrics. Veterinarians and professionals working on cattle health issues are pooled in the World Association for Buiatrics, founded in 1960. National associations and affiliates also exist.

Cattle diseases were in the center of attention in the 1980s and 1990s when the Bovine spongiform encephalopathy (BSE), also known as mad cow disease, was of concern. Cattle might catch and develop various other diseases, like blackleg, bluetongue, foot rot too.

In most states, as cattle health is not only a veterinarian issue, but also a public health issue, public health and food safety standards and farming regulations directly affect the daily work of farmers who keep cattle. However, said rules change frequently and are often debated. For instance, in the U.K., it was proposed in 2011 that milk from tuberculosis-infected cattle should be allowed to enter the food chain. Internal food safety regulations might affect a country's trade policy as well. For example, the United States has just reviewed its beef import rules according to the "mad cow standards"; while Mexico forbids the entry of cattle who are older than 30 months.

Cow urine is commonly used in India for internal medical purposes. It is distilled and then consumed by patients seeking treatment for a wide variety of illnesses. At present, no conclusive medical evidence shows this has any effect. However, an Indian medicine containing cow urine has already obtained U.S. patents.

Digital dermatitis is caused by the bacteria from the genus Treponema. It differs from foot rot and can appear under unsanitary conditions such as poor hygiene or inadequate hoof trimming, among other causes. It primarily affects dairy cattle and has been known to lower the quantity of milk produced, however the milk quality remains unaffected. Cattle are also susceptible to ringworm caused by the fungus, *Trichophyton verrucosum*, a contagious skin disease which may be transferred to humans exposed to infected cows.

Mycobacterium vaccae is a non pathogenic, possibly even beneficial bacteria, that is seen naturally in soil; that was first isolated from cow dung.

Oxen

Oxen (singular ox) are cattle trained as draft animals. Often they are adult, castrated males of larger breeds, although females and bulls are also used in some areas. Usually, an ox is over four years old due to the need for training and to allow it to grow to full size. Oxen are used for plowing, transport, hauling cargo, grain-grinding by trampling or by powering machines, irrigation by powering pumps, and wagon drawing. Oxen were commonly used to skid logs in forests, and sometimes still are, in low-impact, select-cut logging. Oxen are most often used in teams of two, paired, for light work such as carting, with additional pairs added when more power is required, sometimes up to a total of 20 or more.

Draft Zebus in Mumbai, Maharashtra, India

Oxen used in Plowing

An ox is a mature bovine which has learned to respond appropriately to a teamster's signals. These signals are given by verbal commands or by noise (whip cracks). Verbal commands vary according to dialect and local tradition. In one tradition in North America, the commands are:

- "Back up": go backwards

- "Gee": turn right

- "Get up": walk forward

- "Haw": turn left

- "Whoa": stop

Riding an ox in Hova, Sweden

Oxen can pull harder and longer than horses. Though not as fast as horses, they are less prone to injury because they are more sure-footed.

Many oxen are used worldwide, especially in developing countries. About 11.3 million draft oxen are used in sub-Saharan Africa. In India, the number of draft cattle in 1998 was estimated at 65.7 million head. About half the world's crop production is thought to depend on land preparation (such as plowing) made possible by animal traction.

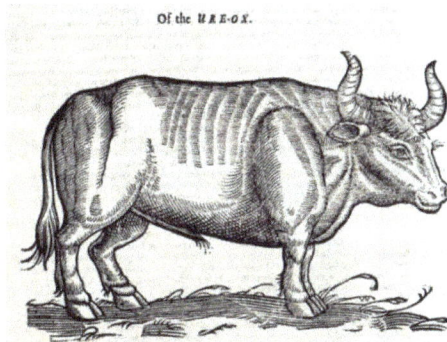
The "Ure-Ox" (Aurochs) by Edward Topsell, 1658

Religion, Traditions and Folklore

Hindu Tradition

Cattle are venerated within the Hindu religion of India. In the Vedic period they were a symbol of plenty and were frequently slaughtered. In later times they gradually acquired their present status. According to the Mahabharata they are to be treated with the same respect 'as one's mother'. In the middle of the first millennium, the consumption of beef began to be disfavoured by lawgivers. Although there has never been any cow-goddesses or temples dedicated to them, cows appear in numerous stories from the Vedas and Puranas. The deity Krishna was brought up in a family of cowherders, and given the name Govinda (protector of the cows). Also, Shiva is traditionally said to ride on the back of a bull named Nandi.

In Hinduism, the cow is a symbol of wealth, strength, abundance, selfless giving and a full Earthly life.

Milk and milk products were used in Vedic rituals. In the postvedic period products of the cow – milk, curd, ghee, but also cow dung and urine (gomutra), or the combination of these five (panchagavya) – began to assume an increasingly important role in ritual purification and expiation.

Veneration of the cow has become a symbol of the identity of Hindus as a community, especially since the end of the 19th century. Slaughter of cows (including oxen, bulls and calves) is forbidden by law in several states of the Indian Union. McDonalds outlets in India do not serve any beef burgers. In Maharaja Ranjit Singh's empire of the early 19th century the killing of a cow was punishable by death.

Other Traditions

Legend of the founding of Durham Cathedral is that monks carrying the body of Saint Cuthbert were led to the location by a milk maid who had lost her dun cow, which was found resting on the spot.

An idealized depiction of girl cow herders in 19th-century Norway by Knud Bergslien.

- The Evangelist St. Luke is depicted as an ox in Christian art.

- In Judaism, as described in Numbers 19:2, the ashes of a sacrificed unblemished red heifer that has never been yoked can be used for ritual purification of people who came into contact with a corpse.

- The ox is one of the 12-year cycle of animals which appear in the Chinese zodiac related to the Chinese calendar.

- The constellation Taurus represents a bull.

- An apocryphal story has it that a cow started the Great Chicago Fire by kicking over a kerosene lamp. Michael Ahern, the reporter who created the cow story, admitted in 1893 that he had fabricated it for more colorful copy.

- On 18 February 1930, Elm Farm Ollie became the first cow to fly in an airplane and also the first cow to be milked in an airplane.

- The first known law requiring branding in North America was enacted on 5 February 1644, by Connecticut. It said that all cattle and pigs had to have a registered brand or earmark by 1 May 1644.

- The akabeko (赤べこ?, *red cow*) is a traditional toy from the Aizu region of Japan that is thought to ward off illness.

- The case of *Sherwood v. Walker*—involving a supposedly barren heifer that was actually pregnant—-first enunciated the concept of mutual mistake as a means of destroying the meeting of the minds in contract law.

- The Fulani of West Africa are the world's largest nomadic cattle-herders.

- The Maasai tribe of East Africa traditionally believe their god Engai entitled them to divine rights to the ownership of all cattle on earth.

Population

For 2013, the FAO estimated global cattle numbers at 1.47 billion. Regionally, the FAO estimate for 2013 includes: Asia 495 million; South America 348 million; Africa 305 million; Europe 122 million; North America 102 million; Central America 46 million; Oceania 42 million; and Caribbean 9 million. The following table shows the cattle population in 2009.

As of 2003, Africa had about 231 million head of cattle, raised in both traditional and non-traditional systems, but often an "integral" part of the culture and way of life.

Cattle population		
Region	**2009**	**2013**
India	285,000,000 (By 2003)	194,655,285

Brazil	187,087,000	186,646,205
China	139,721,000	102,668,900
USA	96,669,000	96,956,461
European Union	87,650,000	
Argentina	51,062,000	52,509,049
Pakistan	38,300,000	26,007,848
Australia	29,202,000	27,249,291
Mexico	26,489,000	31,222,196
Bangladesh	22,976,000	22,844,190
Russian Federation	18,370,000	28,685,315
South Africa	14,187,000	13,526,296
Canada	13,945,000	13,287,866
Other	49,756,000	

References

- Brown, Dave; Sam Meadowcroft (1996). The Modern Shepherd. Wharfedale Road, Ipswich 1P1 4LG, United Kingdom: Farming Press. ISBN 0-85236-188-2.

- Per Jensen (2009). The ethology of domestic animals: an introductory text. CABI. pp. 162–. ISBN 978-1-84593-536-8. Retrieved 15 October 2010.

- Macdonald, David Whyte; Claudio Sillero-Zubiri (2004). The Biology and Conservation of Wild Canids. Oxford University Press. ISBN 0-19-851555-3.

- Ammer, Christine (1997). American Heritage Dictionary of Idioms. Google Books. ISBN 978-0-395-72774-4. Retrieved 2007-11-13.

- Kiple, Kenneth F.; Ornelas, Kriemhild Coneè (2000). The Cambridge World History of Food. Cambridge, UK: Cambridge University Press. ISBN 0-521-40214-X. OCLC 44541840.

- Functional Anatomy and Physiology of Domestic Animals – William O. Reece – Google Boeken. Books.google.com. 4 March 2009. ISBN 978-0-8138-1451-3. Retrieved 2 December 2012.

- G A Slafer – Barley Science: Recent Advances from Molecular Biology to Agronomy of Yield and Quality p.1 Routledge, 12 March 2002 ISBN 1-56022-910-1 Retrieved 2012-06-17

- J Huerta de Soto – 1998 (translated by M.A.Stroup 2012). Money, Bank Credit, and Economic Cycles. Ludwig von Mises Institute. ISBN 1-61016-189-0. Retrieved 2012-06-15.

- "Alaska Isle a Corral For Feral Cattle Herd; U.S. Wants to Trade Cows for Birds". The Washington Post. 2005-10-23. Retrieved 2016-04-26.

- Douglas, Catherine (July 8, 2016), Goats, sheep and cows could challenge dogs for title of 'man's best friend', The Conversation, retrieved 29 August 2016

3

Maintenance of Livestock

The management of farm animals by humans is known as animal husbandry while intensive animal farming means to keep livestock at higher stocking densities than usual. The main purpose of maintaining livestock is to produce eggs, meat and milk for human consumption. This chapter explains the maintenance of livestock by giving a brief description on animal husbandry and intensive animal farming.

Animal Husbandry

Animal husbandry is the management and care of farm animals by humans, in which genetic qualities and behavior, considered to be advantageous to humans, are further developed. The term can refer to the practice of selectively breeding and raising livestock to promote desirable traits in animals for utility, sport, pleasure, or research.

History of Breeding

Animal husbandry has been practiced for thousands of years since the first domestication of animals. Selective breeding for desired traits was first established as a scientific practice by Robert Bakewell during the British Agricultural Revolution in the 18th century. One of his most important breeding programs was with sheep. Using native stock, he was able to quickly select for large, yet fine-boned sheep, with long, lustrous wool. The Lincoln Longwool was improved by Bakewell and in turn the Lincoln was used to develop the subsequent breed, named the New (or Dishley) Leicester. It was hornless and had a square, meaty body with straight top lines. These sheep were exported widely and have contributed to numerous modern breeds.

Under his influence, English farmers began to breed cattle for use primarily as beef for consumption - (previously, cattle were first and foremost bred for pulling ploughs as oxen). Long-horned heifers were crossed with the Westmoreland bull to eventually create the Dishley Longhorn. Over the following decades, farm animals increased dramatically in size and quality. In 1700, the average weight of a bull sold for slaughter was 370 pounds (168 kg). By 1786, that weight had more than doubled to 840 pounds (381 kg).

Animal herding professions specialized in the 19th century to include the cowboy in the United States and Canada, charros and vaqueros in Mexico, gauchos and huasos in South America, and stockmen in Australia.

In more modern times herds are tended on horses, all-terrain vehicles, motorbikes, four-wheel drive vehicles, and helicopters, depending on the terrain and livestock concerned. Today, herd managers often oversee thousands of animals and many staff. Farms, stations and ranches may employ breeders, herd health specialists, feeders, and milkers to help care for the animals.

Breeding Techniques

Techniques such as artificial insemination and embryo transfer are frequently used today, not only as methods to guarantee that females breed regularly but also to help improve herd genetics. This may be done by transplanting embryos from high-quality females into lower-quality surrogate mothers - freeing up the higher-quality mother to be reimpregnated. This practice vastly increases the number of offspring which may be produced by a small selection of the best quality parent animals. On the one hand, this improves the ability of the animals to convert feed to meat, milk, or fiber more efficiently, and improve the quality of the final product. On the other, it decreases genetic diversity, increasing the severity of certain disease outbreaks among other risks.

History in Europe

The semi-natural, unfertilized pastures formed by traditional agricultural methods in Europe, were managed and maintained by the grazing and mowing of livestock. Because the ecological impact of this land management strategy is similar to the impact of a natural disturbance, the agricultural system will share many beneficial characteristics with a natural habitat including the promotion of biodiversity. This strategy is declining in the European context due to the intensification of agriculture, and the mechanized chemical based methods that became popular during and following the industrial revolution.

Good Husbandry Practices

Good husbandry practices (GHP) are a set of rigorous standards whose purpose is to ensure the health of the animals for the production and procurement of products for human consumption.

The checks that are performed within the framework of good husbandry practices tend to reduce the mortality of these animals, and therefore, the spending to cover expenses for illnesses, medications and lost production.

Furthermore, these practices are based on recommendations in order to optimize efficiency in production levels, with high social content that respects the environment and the conditions of individuals who develop tasks related to the agricultural sector.

Factors to Consider for Compliance with GHP

There are elements to consider when organizing production and standardize compliance with good husbandry practices. These criteria allow to optimize the productivity of a farm, based on a previously designed protocol:

- Property Organization

- Number of workers

- Output

- Production Planning

- Facts and history of the hatchery

- Staff Training

Also the GHP set in the protocol the actions to implement for the waste removed from injury, dissection, and milking. The rules addresses methods for separation of solid and liquid waste materials. This intended to prevent blockages in pipes and hoses, in order to limit environmental pollution. Similarly, techniques are expected to separate organic matter colloids (establishment of stabilization ponds). In the same protocol it is important to take into account the costs incurred at each stage and the materials and substances to be used.

Environmental Impact

Animal agriculture is responsible for 20%-33% of all fresh water consumption in the world today.

Livestock or livestock feed occupies 1/3 of the earth's ice-free land. Animal agriculture is the leading cause of species extinction, ocean dead zones, water pollution, and habitat destruction. Animal agriculture contributes to species extinction in many ways. In addition to the monumental habitat destruction caused by clearing forests and converting land to grow feed crops and for animal grazing, predators and "competition" species are frequently targeted and hunted because of a perceived threat to livestock profits. The widespread use of pesticides, herbicides and chemical fertilizers used in the production of feed crops often interferes with the reproductive systems of animals and poison waterways. The overexploitation of wild species through commercial fishing, bushmeat trade as well as animal agriculture's impact on climate change, all contribute to global depletion of species and resources.

Livestock operations on land have created more than 500 nitrogen flooded deadzones around the world in our oceans. Near 1/3 of the planet is desertified, with livestock as the leading driver. A farm with 2,500 dairy cows produces the same amount of waste as a city of 411,000 people.

Animal agriculture is responsible for up to 91% of Amazon destruction.

Climate Change

Due to the significant contribution of agriculture to the emissions of non-CO_2 greenhouse gases, such as methane and nitrous oxide, the relationship between humans and livestock is being analyzed for its potential to help mitigate climate change. Strategies for the mitigation include optimizing the use of gas produced from manure for energy production (biogas).

Livestock and their byproducts account for at least 32,000 million tons of carbon dioxide (CO_2) per year, or 51% of all worldwide greenhouse gas emissions. Livestock is responsible for 65% of all human-related emissions of nitrous oxide – a greenhouse gas with 296 times the global warming potential of carbon dioxide, and which stays in the atmosphere for 150 years. Cows produce an average of 150 billion gallons of methane per day.

Intensive Animal Farming

Intensive animal farming or industrial livestock production, also called factory farming by opponents of the practice, is a modern form of intensive farming that refers to the keeping of livestock, such as cattle, poultry (including in "battery cages") and fish at higher stocking densities than is usually the case with other forms of animal agriculture—a practice typical in industrial farming by agribusinesses. The main products of this industry are meat, milk and eggs for human consumption. There are issues regarding whether factory farming is sustainable and ethical.

Confinement at high stocking density is one part of a systematic effort to produce the highest output at the lowest cost by relying on economies of scale, modern machinery, biotechnology, and global trade. There are differences in the way factory farming techniques are practiced around the world. There is a continuing debate over the benefits, risks and ethical questions of factory farming. The issues include the efficiency of food production; animal welfare; whether it is essential for feeding the growing global population; and the environmental impact (e.g. pollution) and health risks.

History

The practice of industrial animal agriculture is a relatively recent development in the history of agriculture, and the result of scientific discoveries and technological advances.

Innovations in agriculture beginning in the late 19th century generally parallel developments in mass production in other industries that characterized the latter part of the Industrial Revolution. The discovery of vitamins and their role in animal nutrition, in the first two decades of the 20th century, led to vitamin supplements, which allowed chickens to be raised indoors. The discovery of antibiotics and vaccines facilitated raising livestock in larger numbers by reducing disease. Chemicals developed for use in World War II gave rise to synthetic pesticides. Developments in shipping networks and technology have made long-distance distribution of agricultural produce feasible.

Agricultural production across the world doubled four times between 1820 and 1975 (1820 to 1920; 1920 to 1950; 1950 to 1965; and 1965 to 1975) to feed a global population of one billion human beings in 1800 and 6.5 billion in 2002. During the same period, the number of people involved in farming dropped as the process became more automated. In the 1930s, 24 percent of the American population worked in agriculture compared to 1.5 percent in 2002; in 1940, each farm worker supplied 11 consumers, whereas in 2002, each worker supplied 90 consumers.

According to the BBC, the era factory farming per se in Britain began in 1947 when a new Agriculture Act granted subsidies to farmers to encourage greater output by introducing new technology, in order to reduce Britain's reliance on imported meat. The United Nations writes that "intensification of animal production was seen as a way of providing food security." In 1966, the United States, United Kingdom and other industrialized nations, commenced factory farming of Beef and Dariy cattle and domestic pigs. From its American and West European heartland factory farming became globalised in the later years of the 20th century and is still expanding and replacing traditional practices of stock rearing in an increasing number of countries. In 1990 factory farming accounted for 30% of world meat production and by 2005 this had risen to 40%.

Contemporary Animal Production

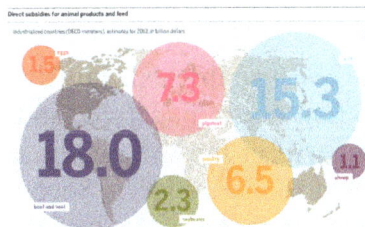

Sum of developed countries' livestock and feed subsidies

Factory farms hold large numbers of animals, typically cows, pigs, turkeys, or chickens, often indoors, typically at high densities. The aim of the operation is to produce large quantities of meat, eggs, or milk at the lowest possible cost. Food is supplied in place. Methods employed to maintain health and improve production may include some combination of disinfectants, antimicrobial agents, anthelmintics, hormones and vaccines; protein, mineral and vitamin supplements; frequent health inspections; biosecurity;

climate-controlled facilities and other measures. Physical restraints, e.g. fences or creeps, are used to control movement or actions regarded as undesirable. Breeding programs are used to produce animals more suited to the confined conditions and able to provide a consistent food product.

Intensive production of livestock and poultry is widespread in developed nations. For 2002-2003, FAO estimates of industrial production as a percentage of global production were 7 percent for beef and veal, 0.8 percent for sheep and goat meat, 42 percent for pork, and 67 percent for poultry meat. Industrial production was estimated to account for 39 percent of the sum of global production of these meats and 50 percent of total egg production. In the U.S., according to its National Pork Producers Council, 80 million of its 95 million pigs slaughtered each year are reared in industrial settings.

Chickens

In the United States, chickens were raised primarily on family farms until 1965. Originally, the primary value in poultry was eggs, and meat was considered a byproduct of egg production. Its supply was less than the demand, and poultry was expensive. Except in hot weather, eggs can be shipped and stored without refrigeration for some time before going bad; this was important in the days before widespread refrigeration.

A commercial chicken house with open sides raising broiler pullets for meat

Farm flocks tended to be small because the hens largely fed themselves through foraging, with some supplementation of grain, scraps, and waste products from other farm ventures. Such feedstuffs were in limited supply, especially in the winter, and this tended to regulate the size of the farm flocks. Soon after poultry keeping gained the attention of agricultural researchers (around 1896), improvements in nutrition and management made poultry keeping more profitable and businesslike.

Prior to about 1910, chicken was served primarily on special occasions or Sunday dinner. Poultry was shipped live or killed, plucked, and packed on ice (but not eviscerated). The "whole, ready-to-cook broiler" was not popular until the 1950s, when end-to-end refrigeration and sanitary practices gave consumers more confidence. Before this, poultry were often cleaned by the neighborhood butcher, though cleaning poultry at home was a commonplace kitchen skill.

Two kinds of poultry were generally used: broilers or "spring chickens"; young male chickens, a byproduct of the egg industry, which were sold when still young and tender (generally under 3 pounds live weight), and "stewing hens", also a byproduct of the egg industry, which were old hens past their prime for laying.

Hens in Brazil

The major milestone in 20th century poultry production was the discovery of vitamin D, which made it possible to keep chickens in confinement year-round. Before this, chickens did not thrive during the winter (due to lack of sunlight), and egg production, incubation, and meat production in the off-season were all very difficult, making poultry a seasonal and expensive proposition. Year-round production lowered costs, especially for broilers.

At the same time, egg production was increased by scientific breeding. After a few false starts, (such as the Maine Experiment Station's failure at improving egg production) success was shown by Professor Dryden at the Oregon Experiment Station.

Improvements in production and quality were accompanied by lower labor requirements. In the 1930s through the early 1950s, 1,500 hens was considered to be a full-time job for a farm family. In the late 1950s, egg prices had fallen so dramatically that farmers typically tripled the number of hens they kept, putting three hens into what had been a single-bird cage or converting their floor-confinement houses from a single deck of roosts to triple-decker roosts. Not long after this, prices fell still further and large numbers of egg farmers left the business.

Robert Plamondon reports that the last family chicken farm in his part of Oregon, Rex Farms, had 30,000 layers and survived into the 1990s. But the standard laying house of the current operators is around 125,000 hens.

This fall in profitability was accompanied by a general fall in prices to the consumer, allowing poultry and eggs to lose their status as luxury foods.

The vertical integration of the egg and poultry industries was a late development, occurring after all the major technological changes had been in place for years (including the development of modern broiler rearing techniques, the adoption of the Cornish Cross broiler, the use of laying cages, etc.).

By the late 1950s, poultry production had changed dramatically. Large farms and packing plants could grow birds by the tens of thousands. Chickens could be sent to slaughterhouses for butchering and processing into prepackaged commercial products to be frozen or shipped fresh to markets or wholesalers. Meat-type chickens currently grow to market weight in six to seven weeks, whereas only fifty years ago it took three times as long. This is due to genetic selection and nutritional modifications (and not the use of growth hormones, which are illegal for use in poultry in the US and many other countries). Once a meat consumed only occasionally, the common availability and lower cost has made chicken a common meat product within developed nations. Growing concerns over the cholesterol content of red meat in the 1980s and 1990s further resulted in increased consumption of chicken.

Today, eggs are produced on large egg ranches on which environmental parameters are well controlled. Chickens are exposed to artificial light cycles to stimulate egg production year-round. In addition, it is a common practice to induce molting through careful manipulation of light and the amount of food they receive in order to further increase egg size and production.

On average, a chicken lays one egg a day, but not on every day of the year. This varies with the breed and time of year. In 1900, average egg production was 83 eggs per hen per year. In 2000, it was well over 300. In the United States, laying hens are butchered after their second egg laying season. In Europe, they are generally butchered after a single season. The laying period begins when the hen is about 18–20 weeks old (depending on breed and season). Males of the egg-type breeds have little commercial value at any age, and all those not used for breeding (roughly fifty percent of all egg-type chickens) are killed soon after hatching. The old hens also have little commercial value. Thus, the main sources of poultry meat 100 years ago (spring chickens and stewing hens) have both been entirely supplanted by meat-type broiler chickens.

Some believe that the "deadly H5N1 strain of bird flu is essentially a problem of industrial poultry practices". On the other hand, according to the CDC article *H5N1 Outbreaks and Enzootic Influenza* by Robert G. Webster et al.:

Transmission of highly pathogenic H5N1 from domestic poultry back to migratory waterfowl in western China has increased the geographic spread. The spread of H5N1 and its likely reintroduction to domestic poultry increase the need for good agricultural vaccines. In fact, the root cause of the continuing H5N1 pandemic threat may be the way the pathogenicity of H5N1 viruses is masked by cocirculating influenza viruses or bad agricultural vaccines.

Webster explains:

If you use a good vaccine you can prevent the transmission within poultry and to humans. But if they have been using vaccines now [in China] for several years, why is there so much bird flu? There is bad vaccine that stops the disease in the bird but the bird goes on pooping out virus and maintaining it and changing it. And I think this is what is going on in China. It has to be. Either there is not enough vaccine being used or there is substandard vaccine being used. Probably both. It's not just China. We can't blame China for substandard vaccines. I think there are substandard vaccines for influenza in poultry all over the world.

In response to the same concerns, Reuters reports Hong Kong infectious disease expert Lo Wing-lok saying that "The issue of vaccines has to take top priority", and Julie Hall, in charge of the WHO's outbreak response in China, saying that China's vaccinations might be "masking" the virus. The BBC reported that Wendy Barclay, a virologist at the University of Reading, UK, said:

The Chinese have made a vaccine based on reverse genetics made with H5N1 antigens, and they have been using it. There has been a lot of criticism of what they have done, because they have protected their chickens against death from this virus but the chickens still get infected; and then you get drift – the virus mutates in response to the antibodies – and now we have a situation where we have five or six "flavours" of H5N1 out there.

Keeping wild birds away from domestic birds is known to be key in the fight against H5N1. Caging (no free range poultry) is one way. Providing wild birds with restored wetlands so they naturally choose nonlivestock areas is another way that helps accomplish this. Political forces are increasingly demanding the selection of one, the other, or both based on nonscientific reasons.

Cattle

Cattle, are domesticated ungulates, a member of the family Bovidae, in the subfamily Bovinae, and descended from the aurochs (*Bos primigenius*). They are raised as livestock for meat (called beef and veal), dairy products (milk), leather and as draught animals (pulling carts, plows and the like). In some countries, such as India, they are honored in religious ceremonies and revered. As of 2009–2010 it is estimated that there are 1.3–1.4 billion head of cattle in the world.

Cattle are often raised by allowing herds to graze on the grasses of large tracts of rangeland called ranches. Raising cattle in this manner allows the productive use of land that might be unsuitable for growing crops. The most common interactions with cattle involve daily feeding, cleaning and milking. Many routine husbandry practices involve ear tagging, dehorning, loading, medical operations, vaccinations and hoof care, as well as training for agricultural shows and preparations. There are also some cultural differences in working with cattle - the cattle husbandry of Fulani men rests on behavioural techniques, whereas in Europe cattle are controlled primarily by physical means like fences.

Once cattle obtain an entry-level weight, about 650 pounds (290 kg), they are transferred from the range to a feedlot to be fed a specialized animal feed which consists of corn byproducts (derived from ethanol production), barley, and other grains as well as alfalfa and cottonseed meal. The feed also contains premixes composed of microingredients such as vitamins, minerals, chemical preservatives, antibiotics, fermentation products, and other essential ingredients that are purchased from premix companies, usually in sacked form, for blending into commercial rations. Because of the availability of these products, a farmer using their own grain can formulate their own rations and be assured the animals are getting the recommended levels of minerals and vitamins. Cattle in the UK are mostly grass fed with the occasional extra such as a mineral lick or feed.

Breeders can utilise cattle husbandry to reduce M. bovis infection susceptibility by selective breeding and maintaining herd health to avoid concurrent disease. Cattle are farmed for beef, veal, dairy, leather and they are sometimes used simply to maintain grassland for wildlife - for example, in Epping Forest, England. They are often used in some of the most wild places for livestock. Depending on the breed, cattle can survive on hill grazing, heaths, marshes, moors and semi desert. Modern cows are more commercial than older breeds and having become more specialised are less versatile. For this reason many smaller farmers still favour old breeds, such as the dairy breed of cattle Jersey.

There are many potential impacts on human health due to the modern cattle industrial agriculture system. There are concerns surrounding the antibiotics and growth hormones used, increased E. Coli contamination, higher saturated fat contents in the meat because of the feed, and also environmental concerns.

As of 2010, in the U.S. 766,350 producers participate in raising beef. The beef industry is segmented with the bulk of the producers participating in raising beef calves. Beef calves are generally raised in small herds, with over 90% of the herds having less than 100 head of cattle. Fewer producers participate in the finishing phase which often occurs in a feedlot, but nonetheless there are 82,170 feedlots in the United States.

Pigs

Intensive piggeries (or hog lots) are a type of concentrated animal feeding operation specialized for the raising of domestic pigs up to slaughterweight. In this system of pig production grower pigs are housed indoors in group-housing or straw-lined sheds, whilst pregnant sows are confined in sow stalls (gestation crates) and give birth in farrowing crates.

The use of sow stalls (gestation crates) has resulted in lower production costs, however, this practice has led to more significant animal welfare concerns. Many of the world's largest producers of pigs (U.S. and Canada) use sow stalls, but some nations (e.g. the UK) and some US States (e.g. Florida and Arizona) have banned them.

Intensive piggeries are generally large warehouse-like buildings. Indoor pig systems allow the pig's condition to be monitored, ensuring minimum fatalities and increased productivity. Buildings are ventilated and their temperature regulated. Most domestic pig varieties are susceptible to heat stress, and all pigs lack sweat glands and cannot cool themselves. Pigs have a limited tolerance to high temperatures and heat stress can lead to death. Maintaining a more specific temperature within the pig-tolerance range also maximizes growth and growth to feed ratio. In an intensive operation pigs will lack access to a wallow (mud), which is their natural cooling mechanism. Intensive piggeries control temperature through ventilation or drip water systems (dropping water to cool the system).

Pigs are naturally omnivorous and are generally fed a combination of grains and protein sources (soybeans, or meat and bone meal). Larger intensive pig farms may be surrounded by farmland where feed-grain crops are grown. Alternatively, piggeries are reliant on the grains industry. Pig feed may be bought packaged or mixed on-site. The intensive piggery system, where pigs are confined in individual stalls, allows each pig to be allotted a portion of feed. The individual feeding system also facilitates individual medication of pigs through feed. This has more significance to intensive farming methods, as the close proximity to other animals enables diseases to spread more rapidly. To prevent disease spreading and encourage growth, drug programs such as antibiotics, vitamins, hormones and other supplements are preemptively administered.

Indoor systems, especially stalls and pens (i.e. 'dry,' not straw-lined systems) allow for the easy collection of waste. In an indoor intensive pig farm, manure can be managed through a lagoon system or other waste-management system. However, odor remains a problem which is difficult to manage.

The way animals are housed in intensive systems varies. Breeding sows will spend the bulk of their time in sow stalls (also called gestation crates) during pregnancy or farrowing crates, with litter, until market.

Piglets often receive range of treatments including castration, tail docking to reduce tail biting, teeth clipped (to reduce injuring their mother's nipples and prevent later tusk growth) and their ears notched to assist identification. Treatments are usually made without pain killers. Weak runts may be slain shortly after birth.

Piglets also may be weaned and removed from the sows at between two and five weeks old and placed in sheds. However, grower pigs - which comprise the bulk of the herd - are usually housed in alternative indoor housing, such as batch pens. During pregnancy, the use of a stall may be preferred as it facilitates feed-management and growth control. It also prevents pig aggression (e.g. tail biting, ear biting, vulva biting, food stealing). Group pens generally require higher stockmanship skills. Such pens will usually not contain straw or other material. Alternatively, a straw-lined shed may house a larger group (i.e. not batched) in age groups.

Many countries have introduced laws to regulate treatment of farmed animals. In the USA, the federal Humane Slaughter Act requires pigs to be stunned before slaughter, although compliance and enforcement is questioned.

Aquaculture

Aquaculture is the cultivation of the natural produce of water (fish, shellfish, algae and other aquatic organisms). The term is distinguished from fishing by the idea of active human effort in maintaining or increasing the number of organisms involved, as opposed to simply taking them from the wild. Subsets of aquaculture include Mariculture (aquaculture in the ocean); Algaculture (the production of kelp/seaweed and other algae); Fish farming (the raising of catfish, tilapia and milkfish in freshwater and brackish ponds or salmon in marine ponds); and the growing of cultured pearls. Extensive aquaculture is based on local photosynthetical production while intensive aquaculture is based on fish fed with an external food supply.

Aquaculture has been used since ancient times and can be found in many cultures. Aquaculture was used in China c. 2500 BC. When the waters lowered after river floods, some fishes, mainly carp, were held in artificial lakes. Their brood were later fed using nymphs and silkworm feces, while the fish themselves were eaten as a source of protein. The Hawaiian people practiced aquaculture by constructing fish ponds. A remarkable example from ancient Hawaii is the construction of a fish pond, dating from at least 1,000 years ago, at Alekoko. The Japanese practiced cultivation of seaweed by providing bamboo poles and, later, nets and oyster shells to serve as anchoring surfaces for spores. The Romans often bred fish in ponds.

The practice of aquaculture gained prevalence in Europe during the Middle Ages, since fish were scarce and thus expensive. However, improvements in transportation during the 19th century made fish easily available and inexpensive, even in inland areas, causing a decline in the practice. The first North American fish hatchery was constructed on Dildo Island, Newfoundland Canada in 1889, it was the largest and most advanced in the world.

Americans were rarely involved in aquaculture until the late 20th century, but California residents harvested wild kelp and made legal efforts to manage the supply starting c. 1900, later even producing it as a wartime resource. (Peter Neushul, Seaweed for War: California's World War I kelp industry, Technology and Culture 30 (July 1989), 561–583)

In contrast to agriculture, the rise of aquaculture is a contemporary phenomenon. According to professor Carlos M. Duarte About 430 (97%) of the aquatic species presently in culture have been domesticated since the start of the 20th century, and an estimated 106 aquatic species have been domesticated over the past decade. The domestication of an aquatic species typically involves about a decade of scientific research. Current success in the domestication of aquatic species results from the 20th century rise of knowledge on the basic biology of aquatic species and the lessons learned from past success and failure. The stagnation in the world's fisheries and overexploitation of 20 to 30% of marine fish species have provided additional impetus to domesticate marine species, just as overexploitation of land animals provided the impetus for the early domestication of land species.

In the 1960s, the price of fish began to climb, as wild fish capture rates peaked and the human population continued to rise. Today, commercial aquaculture exists on an unprecedented, huge scale. In the 1980s, open-netcage salmon farming also expanded; this particular type of aquaculture technology remains a minor part of the production of farmed finfish worldwide, but possible negative impacts on wild stocks, which have come into question since the late 1990s, have caused it to become a major cause of controversy.

In 2003, the total world production of fisheries product was 132.2 million tonnes of which aquaculture contributed 41.9 million tonnes or about 31% of the total world production. The growth rate of worldwide aquaculture is very rapid (greater than 10% per year for most species) while the contribution to the total from wild fisheries has been essentially flat for the last decade.

In the US, approximately 90% of all shrimp consumed are farmed and imported. In recent years salmon aquaculture has become a major export in southern Chile, especially in Puerto Montt and Quellón, Chile's fastest-growing city.

Farmed fish are kept in concentrations never seen in the wild, e.g. 50,000 fish in a 2-acre (8,100 m²) area, with each fish occupying less room than the average bathtub. This can cause several forms of pollution. Packed tightly, fish rub against each other and the sides of their cages, damaging their fins and tails and becoming sickened with various diseases and infections.

Some species of sea lice have been noted to target farmed coho and farmed Atlantic salmon specifically. Such parasites may have an effect on nearby wild fish. For these reasons, aquaculture operators frequently need to use strong drugs to keep the fish alive (but many fish still die prematurely at rates of up to 30%) and these drugs inevitably enter the environment.

The lice and pathogen problems of the 1990s facilitated the development of current treatment methods for sea lice and pathogens. These developments reduced the stress from parasite/pathogen problems. However, being in an ocean environment, the transfer of disease organisms from the wild fish to the aquaculture fish is an ever-present risk factor.

The very large number of fish kept long-term in a single location produces a significant amount of condensed feces, often contaminated with drugs, which again affect local waterways. However, these effects appear to be local to the actual fish farm site and may be minimal to non-measurable in high current sites.

Integrated Multi-trophic Aquaculture

Integrated Multi-Trophic Aquaculture (IMTA) is a practice in which the by-products (wastes) from one species are recycled to become inputs (fertilizers, food) for another. Fed aquaculture (e.g. fish, shrimp) is combined with inorganic extractive (e.g. seaweed) and organic extractive (e.g. shellfish) aquaculture to create balanced systems for environmental sustainability (biomitigation), economic stability (product diversification and risk reduction) and social acceptability (better management practices).

"Multi-Trophic" refers to the incorporation of species from different trophic or nutritional levels in the same system. This is one potential distinction from the age-old practice of aquatic polyculture, which could simply be the co-culture of different fish species from the same trophic level. In this case, these organisms may all share the same biological and chemical processes, with few synergistic benefits, which could potentially lead to significant shifts in the ecosystem. Some traditional polyculture systems may, in fact, incorporate a greater diversity of species, occupying several niches, as extensive cultures (low intensity, low management) within the same pond. The "Integrated" in IMTA refers to the more intensive cultivation of the different species in proximity of each other, connected by nutrient and energy transfer through water, but not necessarily right at the same location.

Ideally, the biological and chemical processes in an IMTA system should balance. This is achieved through the appropriate selection and proportions of different species providing different ecosystem functions. The co-cultured species should be more than just biofilters; they should also be harvestable crops of commercial value. A working IMTA system should result in greater production for the overall system, based on mutual benefits to the co-cultured species and improved ecosystem health, even if the individual production of some of the species is lower compared to what could be reached in monoculture practices over a short term period.

Sometimes the more general term "Integrated Aquaculture" is used to describe the integration of monocultures through water transfer between organisms. For all intents and purposes however, the terms "IMTA" and "integrated aquaculture" differ primarily in their degree of descriptiveness. These terms are sometimes interchanged. Aquaponics, fractionated aquaculture, IAAS (integrated agriculture-aquaculture systems), IPUAS (integrated peri-urban-aquaculture systems), and IFAS (integrated fisheries-aquaculture systems) may also be considered variations of the IMTA concept.

Shrimp

A shrimp farm is an aquaculture business for the cultivation of marine shrimp or prawns for human consumption. Commercial shrimp farming began in the 1970s, and production grew steeply, particularly to match the market demands of the USA, Japan and Western Europe. The total global production of farmed shrimp reached more than

1.6 million tonnes in 2003, representing a value of nearly 9 Billion US$. About 75% of farmed shrimp is produced in Asia, in particular in China and Thailand. The other 25% is produced mainly in Latin America, where Brazil is the largest producer. The largest exporting nation is Thailand.

Shrimp farming has moved from China to Southeast Asia into a meat packing industry. Technological advances have led to growing shrimp at ever higher densities, and broodstock is shipped worldwide. Virtually all farmed shrimp are penaeids (i.e., of the family Penaeidae), and just two species of shrimp—the *Penaeus vannamei* (Pacific white shrimp) and the *Penaeus monodon* (giant tiger prawn)—account for roughly 80% of all farmed shrimp. These industrial monocultures are very susceptible to diseases, which have caused several regional wipe-outs of farm shrimp populations. Increasing ecological problems, repeated disease outbreaks, and pressure and criticism from both NGOs and consumer countries led to changes in the industry in the late 1990s and generally stronger regulation by governments.

Regulation

In various jurisdictions, intensive animal production of some kinds is subject to regulation for environmental protection. In the United States, a CAFO (Concentrated Animal Feeding Operation) that discharges or proposes to discharge waste requires a permit and implementation of a plan for management of manure nutrients, contaminants, wastewater, etc., as applicable, to meet requirements pursuant to the federal Clean Water Act. Some data on regulatory compliance and enforcement are available. In 2000, the US Environmental Protection Agency published 5-year and 1-year data on environmental performance of 32 industries, with data for the livestock industry being derived mostly from inspections of CAFOs. The data pertain to inspections and enforcement mostly under the Clean Water Act, but also under the Clean Air Act and Resource Conservation and Recovery Act. Of the 32 industries, livestock production was among the top seven for environmental performance over the 5-year period, and was one of the top two in the final year of that period, where good environmental performance is indicated by a low ratio of enforcement orders to inspections. The five-year and final-year ratios of enforcement/inspections for the livestock industry were 0.05 and 0.01, respectively. Also in the final year, the livestock industry was one of the two leaders among the 32 industries in terms of having the lowest percentage of facilities with violations. In Canada, intensive livestock operations are subject to provincial regulation, with definitions of regulated entities varying among provinces. Examples include Intensive Livestock Operations (Saskatchewan), Confined Feeding Operations (Alberta), Feedlots (British Columbia), High-density Permanent Outdoor Confinement Areas (Ontario) and Feedlots or Parcs d'Engraissement (Manitoba). In Canada, intensive animal production, like other agricultural sectors, is also subject to various other federal and provincial requirements.

In the United States, farmed animals are excluded by half of all state animal cruelty laws including the federal Animal Welfare Act. The 28-hour law, enacted in 1873 and amended in 1994 states that when animals are being transported for slaughter, the vehicle must stop every 28 hours and the animals must be let out for exercise, food, and water. The United States Department of Agriculture claims that the law does not apply to birds. The Humane Methods of Livestock Slaughter Act is similarly limited. Originally passed in 1958, the Act requires that livestock be stunned into unconsciousness prior to slaughter. This Act also excludes birds, who make up more than 90 percent of the animals slaughtered for food, as well as rabbits and fish. Individual states all have their own animal cruelty statutes; however many states have a provision to exempt standard agricultural practices.

In the United States there is a growing movement to mitigate the worst abuses by regulating factory farming. In Ohio animal welfare organizations reached a negotiated settlement with farm organizations while in California Proposition 2, Standards for Confining Farm Animals, an initiated law was approved by voters in 2008. Regulations have been enacted in other states and plans are underway for referendum and lobbying campaigns in other states.

An action plan has been proposed by the USDA in February 2009, called the Utilization of Manure and Other Agricultural and Industrial Byproducts. This program's goal is to protect the environment and human and animal health by using manure in a safe and effective manner. In order for this to happen, several actions need to be taken and these four components include: • Improving the Usability of Manure Nutrients through More Effective Animal Nutrition and Management • Maximizing the Value of Manure through Improved Collection, Storage, and Treatment Options • Utilizing Manure in Integrated Farming Systems to Improve Profitability and Protect Soil, Water, and Air Quality • Using Manure and Other Agricultural Byproducts as a Renewable Energy Source

In 2012 Australia's largest supermarket chain, Coles, announced that as of January 1, 2013, they will stop selling company branded pork and eggs from animals kept in factory farms. The nation's other dominant supermarket chain, Woolworths, has already begun phasing out factory farmed animal products. All of Woolworth's house brand eggs are now cage-free, and by mid-2013 all of their pork will come from farmers who operate stall-free farms.

Controversies and Criticisms

Advocates of factory farming claim that factory farming has led to the betterment of housing, nutrition, and disease control over the last twenty years, while opponents claim that it harms the environment, creates health risks, and abuses animals.

Animal Welfare

Animal welfare impacts of factory farming can include:

- Close confinement systems (cages, crates) or lifetime confinement in indoor sheds

- Discomfort and injuries caused by inappropriate flooring and housing

- Restriction or prevention of normal exercise and most of natural foraging or exploratory behaviour

- Restriction or prevention of natural maternal nesting behaviour

- Lack of daylight or fresh air and poor air quality in animal sheds

- Social stress and injuries caused by overcrowding

- Health problems caused by extreme selective breeding and management for fast growth and high productivity

- Reduced lifetime (longevity) of breeding animals (dairy cows, breeding sows)

- Fast-spreading infections encouraged by crowding and stress in intensive conditions

- Debeaking (beak trimming or shortening) in the poultry and egg industry to avoid pecking in overcrowded quarters

Confinement and overcrowding of animals results in a lack of exercise and natural locomotory behavior, which weakens their bones and muscles. An intensive poultry farm provides the optimum conditions for viral mutation and transmission – thousands of birds crowded together in a closed, warm, and dusty environment is highly conducive to the transmission of a contagious disease. Selecting generations of birds for their faster growth rates and higher meat yields has left birds' immune systems less able to cope with infections and there is a high degree of genetic uniformity in the population, making the spread of disease more likely. Further intensification of the industry has been suggested by some as the solution to avian flu, on the rationale that keeping birds indoors will prevent contamination. However, this relies on perfect, fail-safe biosecurity – and such measures are near impossible to implement. Movement between farms by people, materials, and vehicles poses a threat and breaches in biosecurity are possible. Intensive farming may be creating highly virulent avian flu strains. With the frequent flow of goods within and between countries, the potential for disease spread is high.

Confinement and overcrowding of animals' environment presents the risk of contamination of the meat from viruses and bacteria. Feedlot animals reside in crowded conditions and often spend their time standing in their own waste. A dairy farm with 2,500 cows may produce as much waste as a city of 411,000 people, and unlike a city in which human waste ends up at a sewage treatment plant, livestock waste is not treated. As a result, feedlot animals have the potential of exposure to various viruses and bacteria via the manure and urine in their environment. Furthermore, the animals often have residual manure on their bodies when they go to slaughter. Sometimes, even "free-range" animals are mutilated without the use of painkillers.

Depending on the kind of system involved, prevention and control of disease in intensive animal farming commonly use (where appropriate) several of biosecurity, sanitation, surveillance, vaccinations, antibiotics, various measures for control of parasites and other pests, preconditioning, low-stress management, and removal of infected animals. According to a February 2011 FDA report, nearly 29 million pounds of antimicrobials were sold in 2009 for both therapeutic and non-therapeutic use for all farm animal species. The Union of Concerned Scientists estimates that 70% of that amount is for non-therapeutic use.

The large concentration of animals, animal waste, and the potential for dead animals in a small space poses ethical issues. It is recognized that some techniques used to sustain intensive agriculture can be cruel to animals such as mutilation. As awareness of the problems of intensive techniques has grown, there have been some efforts by governments and industry to remove inappropriate techniques.

On some farms, chicks may be debeaked when very young, causing pain and shock. Confining hens and pigs in crates no larger than the animal itself may lead to physical problems such as osteoporosis and joint pain, and psychological problems including boredom, depression, and frustration, as shown by repetitive or self-destructive actions. In the UK, the Farm Animal Welfare Council was set up by the government to act as an independent advisor on animal welfare in 1979 and expresses its policy as five freedoms: from hunger & thirst; from discomfort; from pain, injury or disease; to express normal behavior; from fear and distress.

Interior of a gestational sow barn

There are differences around the world as to which practices are accepted and there continue to be changes in regulations with animal welfare being a strong driver for increased regulation. For example, the EU is bringing in further regulation to set maximum stocking densities for meat chickens by 2010, where the UK Animal Welfare Minister commented, "The welfare of meat chickens is a major concern to people throughout the European Union. This agreement sends a strong message to the rest of the world that we care about animal welfare."

Factory farming is greatly debated throughout Australia, with many people disagreeing with the methods and ways in which the animals in factory farms are treated. Animals are often under stress from being kept in confined spaces and will attack each other. In an effort to prevent injury leading to infection, their beaks, tails and teeth are removed. Many piglets will die of shock after having their teeth and tails removed, because pain-killing medicines are not used in these operations. Others say that factory farms are a great way to gain space, with animals such as chickens being kept in spaces smaller than an A4 page.

Less cruel methods of factory farming are still preferable. For example, in the UK, de-beaking of chickens is deprecated, but it is recognized that it is a method of last resort, seen as better than allowing vicious fighting and ultimately cannibalism. Between 60 and 70 percent of six million breeding sows in the U.S. are confined during pregnancy, and for most of their adult lives, in 2 by 7 ft (0.61 by 2.13 m) gestation crates. According to pork producers and many veterinarians, sows will fight if housed in pens. The largest pork producer in the U.S. said in January 2007 that it will phase out gestation crates by 2017. They are being phased out in the European Union, with a ban effective in 2013 after the fourth week of pregnancy. With the evolution of factory farming, there has been a growing awareness of the issues amongst the wider public, not least due to the efforts of animal rights and welfare campaigners. As a result, gestation crates, one of the more contentious practices, are the subject of laws in the U.S., Europe and around the world to phase out their use as a result of pressure to adopt less confined practices.

Human Health Impact

According to the U.S. Centers for Disease Control and Prevention (CDC), farms on which animals are intensively reared can cause adverse health reactions in farm work-ers. Workers may develop acute and chronic lung disease, musculoskeletal injuries, and may catch infections that transmit from animals to human beings (such as tuber-culosis).

Pesticides are used to control organisms which are considered harmful and they save farmers money by preventing product losses to pests. In the US, about a quarter of pesticides used are used in houses, yards, parks, golf courses, and swimming pools and about 70% are used in agriculture. However, pesticides can make their way into con-sumers' bodies which can cause health problems. One source of this is bioaccumulation in animals raised on factory farms.

"Studies have discovered an increase in respiratory, neurobehavioral, and mental ill-nesses among the residents of communities next to factory farms."

The CDC writes that chemical, bacterial, and viral compounds from animal waste may travel in the soil and water. Residents near such farms report problems such as un-pleasant smell, flies and adverse health effects.

The CDC has identified a number of pollutants associated with the discharge of animal waste into rivers and lakes, and into the air. The use of antibiotics may create anti-biotic-resistant pathogens; parasites, bacteria, and viruses may be spread; ammonia, nitrogen, and phosphorus can reduce oxygen in surface waters and contaminate drinking water; pesticides and hormones may cause hormone-related changes in fish; animal feed and feathers may stunt the growth of desirable plants in surface waters and provide nutrients to disease-causing micro-organisms; trace elements such as arsenic and copper, which are harmful to human health, may contaminate surface waters.

Intensive farming may make the evolution and spread of harmful diseases easier. Many communicable animal diseases spread rapidly through densely spaced populations of animals and crowding makes genetic reassortment more likely. However, small family farms are more likely to introduce bird diseases and more frequent association with people into the mix, as happened in the 2009 flu pandemic

In the European Union, growth hormones are banned on the basis that there is no way of determining a safe level. The UK has stated that in the event of the EU raising the ban at some future date, to comply with a precautionary approach, it would only consider the introduction of specific hormones, proven on a case by case basis. In 1998, the European Union banned feeding animals antibiotics that were found to be valuable for human health. Furthermore, in 2006 the European Union banned all drugs for livestock that were used for growth promotion purposes. As a result of these bans, the levels of antibiotic resistance in animal products and within the human population showed a decrease.

The various techniques of factory farming have been associated with a number of European incidents where public health has been threatened or large numbers of animals have had to be slaughtered to deal with disease. Where disease breaks out, it may spread more quickly, not only due to the concentrations of animals, but because modern approaches tend to distribute animals more widely. The international trade in animal products increases the risk of global transmission of virulent diseases such as swine fever, BSE, foot and mouth and bird flu.

In the United States, the use of antibiotics in livestock is still prevalent. The FDA reports that 80 percent of all antibiotics sold in 2009 were administered to livestock animals, and that many of these antibiotics are identical or closely related to drugs used for treating illnesses in humans. Consequently, many of these drugs are losing their effectiveness on humans, and the total healthcare costs associated with drug-resistant bacterial infections in the United States are between $16.6 billion and $26 billion annually.

Methicillin-resistant Staphylococcus aureus (MRSA) has been identified in pigs and humans raising concerns about the role of pigs as reservoirs of MRSA for human infection. One study found that 20% of pig farmers in the United States and Canada in 2007 harbored MRSA. A second study revealed that 81% of Dutch pig farms had pigs with MRSA and 39% of animals at slaughter carried the bug were all of the infections were resistant to tetracycline and many were resistant to other antimicrobials. A more recent study found that MRSA ST398 isolates were less susceptible to tiamulin, an antimicrobial used in agriculture, than other MRSA or methicillin susceptible *S. aureus*. Cases of MRSA have increased in livestock animals. CC398 is a new clone of MRSA that has emerged in animals and is found in intensively reared production animals (primarily pigs, but also cattle and poultry), where it can be transmitted to humans. Although dangerous to humans, CC398 is often asymptomatic in food-producing animals.

A 2011 nationwide study reported nearly half of the meat and poultry sold in U.S. grocery stores — 47 percent — was contaminated with S. aureus, and more than half of those bacteria — 52 percent — were resistant to at least three classes of antibiotics. Although Staph should be killed with proper cooking, it may still pose a risk to consumers through improper food handling and cross-contamination in the kitchen. The senior author of the study said, "The fact that drug-resistant S. aureus was so prevalent, and likely came from the food animals themselves, is troubling, and demands attention to how antibiotics are used in food-animal production today."

In April 2009, lawmakers in the Mexican state of Veracruz accused large-scale hog and poultry operations of being breeding grounds of a pandemic swine flu, although they did not present scientific evidence to support their claim. A swine flu which quickly killed more than 100 infected persons in that area, appears to have begun in the vicinity of a Smithfield subsidiary pig CAFO (concentrated animal feeding operation).

Environmental Impact

Intensive Factory farming has grown to become the biggest threat to the global environment through the loss of ecosystem services and global warming. It is a major driver to global environmental degradation. The process in which feed needs to be grown for animal use only is often grown using intensive methods which involve a significant amount of fertiliser and pesticides. This sometimes results in the pollution of water, soil and air by agrochemicals and manure waste, and use of limited resources such as water and energy at unsustainable rates.

Industrial production of pigs and poultry is an important source of GHG emissions and is predicted to become more so. On intensive pig farms, the animals are generally kept on concrete with slats or grates for the manure to drain through. The manure is usually stored in slurry form (slurry is a liquid mixture of urine and feces). During storage on farm, slurry emits methane and when manure is spread on fields it emits nitrous oxide and causes nitrogen pollution of land and water. Poultry manure from factory farms emits high levels of nitrous oxide and ammonia.

Large quantities and concentrations of waste are produced. Air quality and groundwater are at risk when animal waste is improperly recycled.

Environmental impacts of factory farming include:

- Deforestation for animal feed production

- Unsustainable pressure on land for production of high-protein/high-energy animal feed

- Pesticide, herbicide and fertilizer manufacture and use for feed production

- Unsustainable use of water for feed-crops, including groundwater extraction

- Pollution of soil, water and air by nitrogen and phosphorus from fertiliser used for feed-crops and from manure

- Land degradation (reduced fertility, soil compaction, increased salinity, desertification)

- Loss of biodiversity due to eutrophication, acidification, pesticides and herbicides

- Worldwide reduction of genetic diversity of livestock and loss of traditional breeds

- Species extinctions due to livestock-related habitat destruction (especially feed-cropping)

Labor

Small farmers are often absorbed into factory farm operations, acting as contract growers for the industrial facilities. In the case of poultry contract growers, farmers are required to make costly investments in construction of sheds to house the birds, buy required feed and drugs - often settling for slim profit margins, or even losses.

Market Concentration

The major concentration of the industry occurs at the slaughter and meat processing phase, with only four companies slaughtering and processing 81 percent of cows, 73 percent of sheep, 57 percent of pigs and 50 percent of chickens. This concentration at the slaughter phase may be in large part due to regulatory barriers that may make it financially difficult for small slaughter plants to be built, maintained or remain in business. Factory farming may be no more beneficial to livestock producers than traditional farming because it appears to contribute to overproduction that drives down prices. Through "forward contracts" and "marketing agreements", meatpackers are able to set the price of livestock long before they are ready for production. These strategies often cause farmers to lose money, as half of all U.S. family farming operations did in 2007.

In 1967, there were one million pig farms in America; as of 2002, there were 114,000.

Many of the nation's livestock producers would like to market livestock directly to consumers but with limited USDA inspected slaughter facilities, livestock grown locally can not typically be slaughtered and processed locally.

Demonstrations

From 2011 to 2014 each year between 15,000 and 30,000 people gathered under the theme *We are fed up!* in Berlin to protest against industrial livestock production.

References

- Duram, Leslie A. (2010). Encyclopedia of Organic, Sustainable, and Local Food. ABC-CLIO. p. 139. ISBN 0-313-35963-6.

- "Health and Consumer Protection - Scientific Committee on Animal Health and Animal Welfare - Previous outcome of discussions (Scientific Veterinary Committee) - 17". Retrieved September 6, 2015.

- Food Standards Agency. "[ARCHIVED CONTENT] Food Standards Agency - VPC report on growth hormones in meat". Retrieved September 6, 2015.

- "US meat and poultry is widely contaminated with drug-resistant Staph bacteria, study finds". Retrieved September 6, 2015.

- "Farmers defend themselves at Berlin's Green Week - Business - DW.COM - 17.01.2014". DW.COM. Retrieved September 6, 2015.

- "Mercy For Animals – World's Leading Farmed Animal Rights and Vegan Advocacy Organization - Mercy For Animals". Mercy For Animals. December 17, 2014. Retrieved September 6, 2015.

- "Factory Farm Nation: How America Turned its Livestock Farms into Factories" (PDF). November 2010. Retrieved July 22, 2012.

4

Commodities Produced by Livestock

Humans, for a number of reasons and purposes use livestock. Livestock products include fiber, fur, honey, dairy products etc. Modern techniques seek to minimize human involvement while increasing outcome and improving animal health. The following content helps the reader to develop a better understanding on the commodities produced by livestock.

Food

Food is any substance consumed to provide nutritional support for the body. It is usually of plant or animal origin, and contains essential nutrients, such as fats, proteins, vitamins, or minerals. The substance is ingested by an organism and assimilated by the organism's cells to provide energy, maintain life, or stimulate growth.

Historically, people secured food through two methods: hunting and gathering and agriculture. Today, the majority of the food energy required by the ever increasing population of the world is supplied by the food industry.

Food safety and food security are monitored by agencies like the International Association for Food Protection, World Resources Institute, World Food Programme, Food and Agriculture Organization, and International Food Information Council. They address issues such as sustainability, biological diversity, climate change, nutritional economics, population growth, water supply, and access to food.

The right to food is a human right derived from the International Covenant on Economic, Social and Cultural Rights (ICESCR), recognizing the "right to an adequate standard of living, including adequate food", as well as the "fundamental right to be free from hunger".

Various foods

Foods from plant sources

Food Sources

Most food has its origin in plants. Some food is obtained directly from plants; but even animals that are used as food sources are raised by feeding them food derived from plants. Cereal grain is a staple food that provides more food energy worldwide than any other type of crop. Corn (maize), wheat, and rice – in all of their varieties – account for 87% of all grain production worldwide. Most of the grain that is produced worldwide is fed to livestock.

Some foods not from animal or plant sources include various edible fungi, especially mushrooms. Fungi and ambient bacteria are used in the preparation of fermented and pickled foods like leavened bread, alcoholic drinks, cheese, pickles, kombucha, and yogurt. Another example is blue-green algae such as Spirulina. Inorganic substances such as salt, baking soda and cream of tartar are used to preserve or chemically alter an ingredient.

Plants

Many plants and plant parts are eaten as food and around 2,000 plant species are cultivated for food. Many of these plant species have several distinct cultivars.

Seeds of plants are a good source of food for animals, including humans, because they contain the nutrients necessary for the plant's initial growth, including many healthful fats, such as Omega fats. In fact, the majority of food consumed by human beings are seed-based foods. Edible seeds include cereals (corn, wheat, rice, et cetera), legumes (beans, peas, lentils, et cetera), and nuts. Oilseeds are often pressed to produce rich oils - sunflower, flaxseed, rapeseed (including canola oil), sesame, et cetera.

Seeds are typically high in unsaturated fats and, in moderation, are considered a health food, although not all seeds are edible. Large seeds, such as those from a lemon, pose a choking hazard, while seeds from cherries and apples contain cyanide which could be poisonous only if consumed in large volumes.

Fruits are the ripened ovaries of plants, including the seeds within. Many plants and animals have coevolved such that the fruits of the former are an attractive food source to the latter, because animals that eat the fruits may excrete the seeds some distance away. Fruits, therefore, make up a significant part of the diets of most cultures. Some botanical fruits, such as tomatoes, pumpkins, and eggplants, are eaten as vegetables.

Vegetables are a second type of plant matter that is commonly eaten as food. These include root vegetables (potatoes and carrots), bulbs (onion family), leaf vegetables (spinach and lettuce), stem vegetables (bamboo shoots and asparagus), and inflorescence vegetables (globe artichokes and broccoli and other vegetables such as cabbage or cauliflower).

Animals

Animals are used as food either directly or indirectly by the products they produce. Meat is an example of a direct product taken from an animal, which comes from muscle systems or from organs.

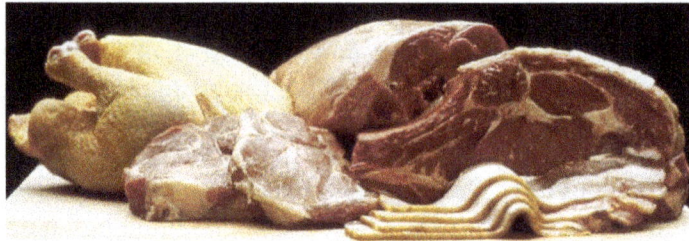
Various raw meats

Food products produced by animals include milk produced by mammary glands, which in many cultures is drunk or processed into dairy products (cheese, butter, etc.). In addition, birds and other animals lay eggs, which are often eaten, and bees produce honey, a reduced nectar from flowers, which is a popular sweetener in many cultures. Some cultures consume blood, sometimes in the form of blood sausage, as a thickener for sauces, or in a cured, salted form for times of food scarcity, and others use blood in stews such as jugged hare.

Some cultures and people do not consume meat or animal food products for cultural, dietary, health, ethical, or ideological reasons. Vegetarians choose to forgo food from animal sources to varying degrees. Vegans do not consume any foods that are or contain ingredients from an animal source.

Production

Most food has always been obtained through agriculture. With increasing concern over both the methods and products of modern industrial agriculture, there has been a growing trend toward sustainable agricultural practices. This approach, partly fueled by consumer demand, encourages biodiversity, local self-reliance and organic farming methods. Major influences on food production include international organizations (e.g. the World Trade Organization and Common Agricultural Policy), national government policy (or law), and war.

A tractor and chaser bin

In popular culture, the mass production of food, specifically meats such as chicken and beef, has come under fire from various documentaries, most recently Food, Inc, documenting the mass slaughter and poor treatment of animals, often for easier revenues from large corporations. Along with a current trend towards environmentalism, people in Western culture have had an increasing trend towards the use of herbal supplements, foods for a specific group of people (such as dieters, women, or athletes), functional foods (fortified foods, such as omega-3 eggs), and a more ethnically diverse diet.

Several organisations have begun calling for a new kind of agriculture in which agroecosystems provide food but also support vital ecosystem services so that soil fertility and biodiversity are maintained rather than compromised. According to the International Water Management Institute and UNEP, well-managed agroecosystems not only provide food, fiber and animal products, they also provide services such as flood mitigation, groundwater recharge, erosion control and habitats for plants, birds fish and other animals.

Taste Perception

Animals, specifically humans, have five different types of tastes: sweet, sour, salty, bitter, and umami. As animals have evolved, the tastes that provide the most energy (sugar and fats) are the most pleasant to eat while others, such as bitter, are not enjoyable. Water, while important for survival, has no taste. Fats, on the other hand, especially saturated fats, are thicker and rich and are thus considered more enjoyable to eat.

Sweet

Generally regarded as the most pleasant taste, sweetness is almost always caused by a type of simple sugar such as glucose or fructose, or disaccharides such as sucrose, a molecule combining glucose and fructose. Complex carbohydrates are long chains and thus do not have the sweet taste. Artificial sweeteners such as sucralose are used to mimic the sugar molecule, creating the sensation of sweet, without the calories. Other types of sugar include raw sugar, which is known for its amber color, as it is unprocessed. As sugar is vital for energy and survival, the taste of sugar is pleasant.

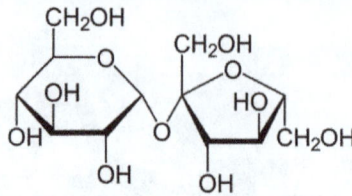

Structure of sucrose

The stevia plant contains a compound known as steviol which, when extracted, has 300 times the sweetness of sugar while having minimal impact on blood sugar.

Sour

Sourness is caused by the taste of acids, such as vinegar in alcoholic beverages. Sour foods include citrus, specifically lemons, limes, and to a lesser degree oranges. Sour is evolutionarily significant as it is a sign for a food that may have gone rancid due to bacteria. Many foods, however, are slightly acidic, and help stimulate the taste buds and enhance flavor.

Salty

Saltiness is the taste of alkali metal ions such as sodium and potassium. It is found in almost every food in low to moderate proportions to enhance flavor, although to eat pure salt is regarded as highly unpleasant. There are many different types of salt, with each having a different degree of saltiness, including sea salt, fleur de sel, kosher salt, mined salt, and grey salt. Other than enhancing flavor, its significance is that the body needs and maintains a delicate electrolyte balance, which is the kidney's function. Salt may be iodized, meaning iodine has been added to it, a necessary nutrient that promotes thyroid function. Some canned foods, notably soups or packaged broths, tend to be high in salt as a means of preserving the food longer. Historically salt has long been used as a meat preservative as salt promotes water excretion. Similarly, dried foods also promote food safety.

Salt mounds in Bolivia

Bitter

Bitterness is a sensation often considered unpleasant characterized by having a sharp, pungent taste. Unsweetened dark chocolate, caffeine, lemon rind, and some types of fruit are known to be bitter.

Umami

Umami, the Japanese word for delicious, is the least known in Western popular culture but has a long tradition in Asian cuisine. Umami is the taste of glutamates, especially monosodium glutamate (MSG). It is characterized as savory, meaty, and rich in flavor. Salmon and mushrooms are foods high in umami.

Cuisine

Many cultures have a recognizable cuisine, a specific set of cooking traditions using various spices or a combination of flavors unique to that culture, which evolves over time. Other differences include preferences (hot or cold, spicy, etc.) and practices, the study of which is known as gastronomy. Many cultures have diversified their foods by means of preparation, cooking methods, and manufacturing. This also includes a complex food trade which helps the cultures to economically survive by way of food, not just by consumption.

Typical Assyrian cuisine

Some popular types of ethnic foods include Italian, French, Japanese, Chinese, American, Cajun, Thai, African, and Indian cuisine. Various cultures throughout the world study the dietary analysis of food habits. While evolutionarily speaking, as opposed to culturally, humans are omnivores, religion and social constructs such as morality, activism, or environmentalism will often affect which foods they will consume. Food is eaten and typically enjoyed through the sense of taste, the perception of flavor from eating and drinking. Certain tastes are more enjoyable than others, for evolutionary purposes.

Presentation

Aesthetically pleasing and eye-appealing food presentations can encourage people to consume foods. A common saying is that people "eat with their eyes". Food presented in a clean and appetizing way will encourage a good flavor, even if unsatisfactory.

A French basil salmon terrine, with eye-appealing garnishes

Contrast in Texture

Texture plays a crucial role in the enjoyment of eating foods. Contrasts in textures, such as something crunchy in an otherwise smooth dish, may increase the appeal of eating it. Common examples include adding granola to yogurt, adding croutons to a salad or soup, and toasting bread to enhance its crunchiness for a smooth topping, such as jam or butter.

Contrast in Taste

Another universal phenomenon regarding food is the appeal of contrast in taste and presentation. For example, such opposite flavors as sweetness and saltiness tend to go well together, as in kettle corn and nuts.

Food Preparation

While many foods can be eaten raw, many also undergo some form of preparation for reasons of safety, palatability, texture, or flavor. At the simplest level this may involve washing, cutting, trimming, or adding other foods or ingredients, such as spices. It may also involve mixing, heating or cooling, pressure cooking, fermentation, or combination with other food. In a home, most food preparation takes place in a kitchen. Some preparation is done to enhance the taste or aesthetic appeal; other preparation may help to preserve the food; others may be involved in cultural identity. A meal is made up of food which is prepared to be eaten at a specific time and place.

A refrigerator helps to keep foods fresh.

Animal Preparation

The preparation of animal-based food usually involves slaughter, evisceration, hanging, portioning, and rendering. In developed countries, this is usually done outside the home in slaughterhouses, which are used to process animals en masse for meat production. Many countries regulate their slaughterhouses by law. For example, the United States has established the Humane Slaughter Act of 1958, which requires that an animal be stunned before killing. This act, like those in many countries, exempts slaughter in accordance to religious law, such as kosher, shechita, and dhabiha halal. Strict interpretations of kashrut require the animal to be fully aware when its carotid artery is cut.

On the local level, a butcher may commonly break down larger animal meat into smaller manageable cuts, and pre-wrap them for commercial sale or wrap them to order in butcher paper. In addition, fish and seafood may be fabricated into smaller cuts by a fish monger. However fish butchery may be done on board a fishing vessel and quick-frozen for preservation of quality.

Cooking

The term "cooking" encompasses a vast range of methods, tools, and combinations of ingredients to improve the flavor or digestibility of food. Cooking technique, known as culinary art, generally requires the selection, measurement, and combining of ingredients in an ordered procedure in an effort to achieve the desired result. Constraints on success include the variability of ingredients, ambient conditions, tools, and the skill of the individual cook. The diversity of cooking worldwide is a reflection of the myriad nutritional, aesthetic, agricultural, economic, cultural, and religious considerations that affect it.

Cooking with a wok in China

Cooking requires applying heat to a food which usually, though not always, chemically changes the molecules, thus changing its flavor, texture, appearance, and nutritional properties. Cooking certain proteins, such as egg whites, meats, and fish, denatures the protein, causing it to firm. There is archaeological evidence of roasted foodstuffs at *Homo erectus* campsites dating from 420,000 years ago. Boiling as a means of cooking requires a container, and has been practiced at least since the 10th millennium BC with the introduction of pottery.

Cooking Equipment

There are many different types of equipment used for cooking.

A stainless steel frying pan

A traditional asado (barbecue)

Ovens are mostly hollow devices that get very hot (up to 500 °F) and are used for baking or roasting and offer a dry-heat cooking method. Different cuisines will use different types of ovens. For example, Indian culture uses a Tandoor oven, which is a cylindrical clay oven which operates at a single high temperature. Western kitchens use variable temperature convection ovens, conventional ovens, toaster ovens, or non-radiant heat ovens like the microwave oven. Classic Italian cuisine includes the use of a brick oven containing burning wood. Ovens may be wood-fired, coal-fired, gas, electric, or oil-fired.

Various types of cook-tops are used as well. They carry the same variations of fuel types as the ovens mentioned above. Cook-tops are used to heat vessels placed on top of the heat source, such as a sauté pan, sauce pot, frying pan, or pressure cooker. These pieces of equipment can use either a moist or dry cooking method and include methods such as steaming, simmering, boiling, and poaching for moist methods, while the dry methods include sautéing, pan frying, and deep-frying.

In addition, many cultures use grills for cooking. A grill operates with a radiant heat source from below, usually covered with a metal grid and sometimes a cover. An open pit barbecue in the American south is one example along with the American style outdoor grill fueled by wood, liquid propane, or charcoal along with soaked wood chips for smoking. A Mexican style of barbecue is called barbacoa, which involves the cooking of meats such as whole sheep over an open fire. In Argentina, an asado (Spanish for "grilled") is prepared on a grill held over an open pit or fire made upon the ground, on which a whole animal or smaller cuts are grilled.

Raw Food Preparation

Certain cultures highlight animal and vegetable foods in their raw state. Salads consisting of raw vegetables or fruits are common in many cuisines. Sashimi in Japanese cuisine consists of raw sliced fish or other meat, and sushi often incorporates raw fish or seafood. Steak tartare and salmon tartare are dishes made from diced or ground raw beef or salmon, mixed with various ingredients and served with baguettes, brioche, or frites. In Italy, carpaccio is a dish of very thinly sliced raw beef, drizzled with a vinaigrette made with olive oil. The health food movement known as raw foodism promotes a mostly vegan diet of raw fruits, vegetables, and grains prepared in various ways, including juicing, food dehydration, sprouting, and other methods of preparation that do not heat the food above 118 °F (47.8 °C). An example of a raw meat dish is ceviche, a Latin American dish made with raw meat that is "cooked" from the highly acidic citric juice from lemons and limes along with other aromatics such as garlic.

Many types of fish ready to be eaten, including salmon and tuna

Restaurants

Restaurants employ trained chefs who prepare food, and trained waitstaff to serve the customers. The term restaurant is credited to the French from the 19th century, as it relates to the restorative nature of the bouillons that were once served in them. However, the concept pre-dates the naming of these establishments, as evidence suggests commercial food preparation may have existed during the age of the city of Pompeii, and urban sales of prepared foods may have existed in China during the Song dynasty. The coffee shops or cafés of 17th century Europe may also be considered an early version of the restaurant. In 2005, the population of the United States spent $496 billion for out-of-home dining. Expenditures by type of out-of-home dining were as follows:

40% in full-service restaurants, 37.2% in limited service restaurants (fast food), 6.6% in schools or colleges, 5.4% in bars and vending machines, 4.7% in hotels and motels, 4.0% in recreational places, and 2.2% in others, which includes military bases.

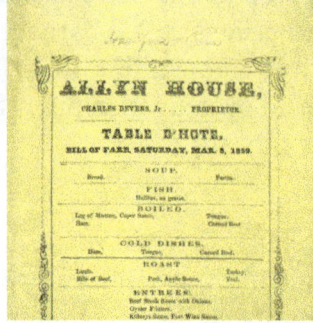

The Allyn House restaurant menu (March 5, 1859)

A McDonald's restaurant in Riyadh, Saudi Arabia

Tom's Restaurant, a restaurant in New York City

Food Manufacturing

Packaged foods are manufactured outside the home for purchase. This can be as simple as a butcher preparing meat, or as complex as a modern international food industry. Early food processing techniques were limited by available food preservation, packaging, and transportation. This mainly involved salting, curing, curdling, drying, pickling, fermenting, and smoking. Food manufacturing arose during the industrial revolution in the 19th century. This development took advantage of new mass markets and emerging technology, such as milling, preservation, packaging and labeling, and transportation. It brought the advantages of pre-prepared time-saving food to the bulk of ordinary people who did not employ domestic servants.

Packaged household food items

At the start of the 21st century, a two-tier structure has arisen, with a few international food processing giants controlling a wide range of well-known food brands. There also exists a wide array of small local or national food processing companies. Advanced technologies have also come to change food manufacture. Computer-based control systems, sophisticated processing and packaging methods, and logistics and distribution advances can enhance product quality, improve food safety, and reduce costs.

Commercial Trade

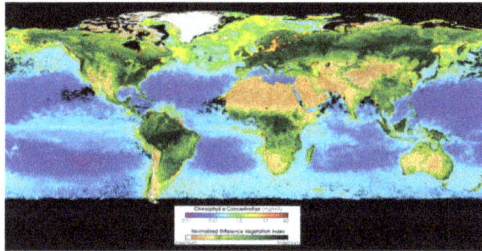

SeaWiFS image for the global biosphere

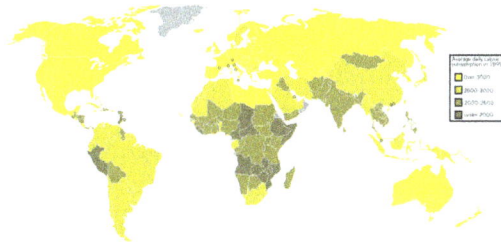

Global average daily calorie consumption in 1995

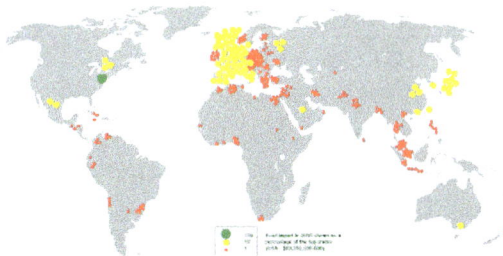

Food imports in 2005

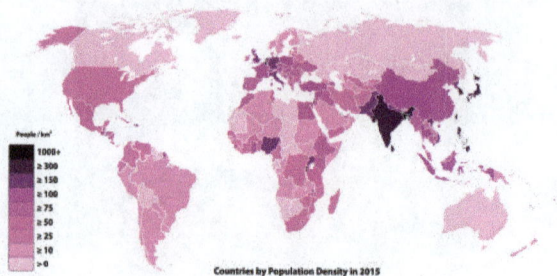

Population density of world regions

International Food Imports and Exports

The World Bank reported that the European Union was the top food importer in 2005, followed at a distance by the USA and Japan. Britain's need for food was especially well illustrated in World War II. Despite the implementation of food rationing, Britain remained dependent on food imports and the result was a long term engagement in the Battle of the Atlantic.

Food is traded and marketed on a global basis. The variety and availability of food is no longer restricted by the diversity of locally grown food or the limitations of the local growing season. Between 1961 and 1999, there was a 400% increase in worldwide food exports. Some countries are now economically dependent on food exports, which in some cases account for over 80% of all exports.

In 1994, over 100 countries became signatories to the Uruguay Round of the General Agreement on Tariffs and Trade in a dramatic increase in trade liberalization. This included an agreement to reduce subsidies paid to farmers, underpinned by the WTO enforcement of agricultural subsidy, tariffs, import quotas, and settlement of trade disputes that cannot be bilaterally resolved. Where trade barriers are raised on the disputed grounds of public health and safety, the WTO refer the dispute to the Codex Alimentarius Commission, which was founded in 1962 by the United Nations Food and Agriculture Organization and the World Health Organization. Trade liberalization has greatly affected world food trade.

Marketing and Retailing

Food marketing brings together the producer and the consumer. It is the chain of activities that brings food from "farm gate to plate". The marketing of even a single food product can be a complicated process involving many producers and companies. For example, fifty-six companies are involved in making one can of chicken noodle soup. These businesses include not only chicken and vegetable processors but also the companies that transport the ingredients and those who print labels and manufacture cans. The food marketing system is the largest direct and indirect non-government employer in the United States.

Packaged food aisles of supermarket in Portland, Oregon, United States of America

In the pre-modern era, the sale of surplus food took place once a week when farmers took their wares on market day into the local village marketplace. Here food was sold to grocers for sale in their local shops for purchase by local consumers. With the onset of industrialization and the development of the food processing industry, a wider range of food could be sold and distributed in distant locations. Typically early grocery shops would be counter-based shops, in which purchasers told the shop-keeper what they wanted, so that the shop-keeper could get it for them.

In the 20th century, supermarkets were born. Supermarkets brought with them a self ser-vice approach to shopping using shopping carts, and were able to offer quality food at lower cost through economies of scale and reduced staffing costs. In the latter part of the 20th century, this has been further revolutionized by the development of vast warehouse-sized, out-of-town supermarkets, selling a wide range of food from around the world.

Unlike food processors, food retailing is a two-tier market in which a small number of very large companies control a large proportion of supermarkets. The supermarket giants wield great purchasing power over farmers and processors, and strong influence over consumers. Nevertheless, less than 10% of consumer spending on food goes to farmers, with larger percentages going to advertising, transportation, and intermediate corporations.

Prices

Some essential food products including bread, rice and pasta

It was reported on March 24, 2008, that consumers worldwide faced rising food prices. Reasons for this development include changes in the weather and dramatic changes in the global economy, including higher oil prices, lower food reserves, and growing consumer demand in China and India. In the long term, prices are expected to stabilize. Farmers will grow more grain for both fuel and food and eventually bring prices down. Already this is happening with wheat, with more crops to be planted in the United States, Canada, and Europe in 2009. However, the Food and Agriculture Organization projects that consumers still have to deal with more expensive food until at least 2018.

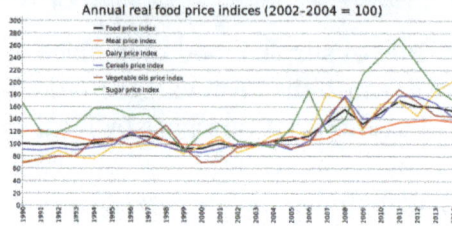

Annual real food price indices (2002-2004 = 100)

Food, meat, dairy, cereals, vegetable oil, and sugar price indices, deflated using the World Bank Manufactures Unit Value Index (MUV)

It is rare for the spikes to hit all major foods in most countries at once. Food prices rose 4% in the United States in 2007, the highest increase since 1990, and are expected to climb as much again in 2008. As of December 2007, 37 countries faced food crises, and 20 had imposed some sort of food-price controls. In China, the price of pork jumped 58% in 2007. In the 1980s and 1990s, farm subsidies and support programs allowed major grain exporting countries to hold large surpluses, which could be tapped during food shortages to keep prices down. However, new trade policies have made agricultural production much more responsive to market demands, putting global food reserves at their lowest since 1983.

Food prices are rising, wealthier Asian consumers are westernizing their diets, and farmers and nations of the third world are struggling to keep up the pace. The past five years have seen rapid growth in the contribution of Asian nations to the global fluid and powdered milk manufacturing industry, which in 2008 accounted for more than 30% of production, while China alone accounts for more than 10% of both production and consumption in the global fruit and vegetable processing and preserving industry. The trend is similarly evident in industries such as soft drink and bottled water manufacturing, as well as global cocoa, chocolate, and sugar confectionery manufacturing, forecast to grow by 5.7% and 10.0% respectively during 2008 in response to soaring demand in Chinese and Southeast Asian markets.

Rising food prices over recent years have been linked with social unrest around the world, including rioting in Bangladesh and Mexico, and the Arab Spring.

In 2013 Overseas Development Institute researchers showed that rice has more than doubled in price since 2000, rising by 120% in real terms. This was as a result of shifts in trade policy and restocking by major producers. More fundamental drivers of increased prices are the higher costs of fertiliser, diesel and labour. Parts of Asia see rural wages rise with potential large benefits for the 1.3 billion (2008 estimate) of Asia's poor in reducing the poverty they face. However, this negatively impacts more vulnerable groups who don't share in the economic boom, especially in Asian and African coastal cities. The researchers said the threat means social-protection policies are needed to guard against price shocks. The research proposed that in the longer run, the rises present opportunities to export for Western African farmers with high potential for rice production to replace imports with domestic production.

As Investment

Institutions such as hedge funds, pension funds and investment banks like Barclays Capital, Goldman Sachs and Morgan Stanley have been instrumental in pushing up prices in the last five years, with investment in food commodities rising from $65bn to $126bn (£41bn to £79bn) between 2007 and 2012, contributing to 30-year highs. This has caused price fluctuations which are not strongly related to the actual supply of food, according to the United Nations. Financial institutions now make up 61% of all investment in wheat futures. According to Olivier De Schutter, the UN special rapporteur on food, there was a rush by institutions to enter the food market following George W Bush's Commodities Futures Modernization Act of 2000. De Schutter told the Independent in March 2012: "What we are seeing now is that these financial markets have developed massively with the arrival of these new financial investors, who are purely interested in the short-term monetary gain and are not really interested in the physical thing – they never actually buy the ton of wheat or maize; they only buy a promise to buy or to sell. The result of this financialisation of the commodities market is that the prices of the products respond increasingly to a purely speculative logic. This explains why in very short periods of time we see prices spiking or bubbles exploding, because prices are less and less determined by the real match between supply and demand." In 2011, 450 economists from around the world called on the G20 to regulate the commodities market more.

Some experts have said that speculation has merely aggravated other factors, such as climate change, competition with bio-fuels and overall rising demand. However, some such as Jayati Ghosh, professor of economics at Jawaharlal Nehru University in New Delhi, have pointed out that prices have increased irrespective of supply and demand issues: Ghosh points to world wheat prices, which doubled in the period from June to December 2010, despite there being no fall in global supply.

Famine and Hunger

Food deprivation leads to malnutrition and ultimately starvation. This is often connected with famine, which involves the absence of food in entire communities. This can have a devastating and widespread effect on human health and mortality. Rationing is sometimes used to distribute food in times of shortage, most notably during times of war.

Starvation is a significant international problem. Approximately 815 million people are undernourished, and over 16,000 children die per day from hunger-related causes. Food deprivation is regarded as a deficit need in Maslow's hierarchy of needs and is measured using famine scales.

Food Aid

Food aid can benefit people suffering from a shortage of food. It can be used to improve peoples' lives in the short term, so that a society can increase its standard of living to the point that food aid is no longer required. Conversely, badly managed food aid can create problems by disrupting local markets, depressing crop prices, and discouraging food production. Sometimes a cycle of food aid dependence can develop. Its provision, or threatened withdrawal, is sometimes used as a political tool to influence the policies of the destination country, a strategy known as food politics. Sometimes, food aid provisions will require certain types of food be purchased from certain sellers, and food aid can be misused to enhance the markets of donor countries. International efforts to distribute food to the neediest countries are often coordinated by the World Food Programme.

Safety

Foodborne illness, commonly called "food poisoning", is caused by bacteria, toxins, viruses, parasites, and prions. Roughly 7 million people die of food poisoning each year, with about 10 times as many suffering from a non-fatal version. The two most common factors leading to cases of bacterial foodborne illness are cross-contamination of ready-to-eat food from other uncooked foods and improper temperature control. Less commonly, acute adverse reactions can also occur if chemical contamination of food occurs, for example from improper storage, or use of non-food grade soaps and disinfectants. Food can also be adulterated by a very wide range of articles (known as "foreign bodies") during farming, manufacture, cooking, packaging, distribution, or sale. These foreign bodies can include pests or their droppings, hairs, cigarette butts, wood chips, and all manner of other contaminants. It is possible for certain types of food to become contaminated if stored or presented in an unsafe container, such as a ceramic pot with lead-based glaze.

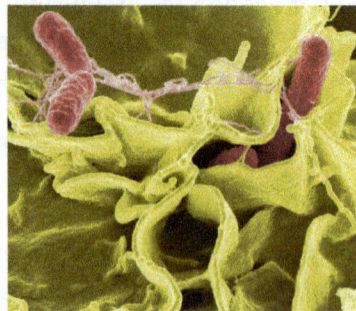

Salmonella bacteria is a common cause of foodborne illness, particularly in undercooked chicken and chicken eggs.

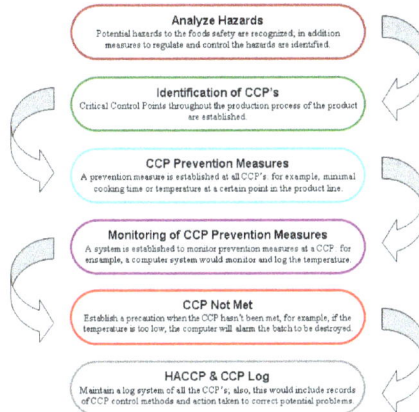

Hazard Analysis and Critical Control Points (HACCP) Flowchart

Food poisoning has been recognized as a disease since as early as Hippocrates. The sale of rancid, contaminated, or adulterated food was commonplace until the introduction of hygiene, refrigeration, and vermin controls in the 19th century. Discovery of techniques for killing bacteria using heat, and other microbiological studies by scientists such as Louis Pasteur, contributed to the modern sanitation standards that are ubiquitous in developed nations today. This was further underpinned by the work of Justus von Liebig, which led to the development of modern food storage and food preservation methods. In more recent years, a greater understanding of the causes of food-borne illnesses has led to the development of more systematic approaches such as the Hazard Analysis and Critical Control Points (HACCP), which can identify and eliminate many risks.

Recommended measures for ensuring food safety include maintaining a clean preparation area with foods of different types kept separate, ensuring an adequate cooking temperature, and refrigerating foods promptly after cooking.

Foods that spoil easily, such as meats, dairy, and seafood, must be prepared a certain way to avoid contaminating the people for whom they are prepared. As such, the rule of thumb is that cold foods (such as dairy products) should be kept cold and hot foods (such as soup) should be kept hot until storage. Cold meats, such as chicken, that are to be cooked should not be placed at room temperature for thawing, at the risk of dangerous bacterial growth, such as *Salmonella* or *E. coli*.

Allergies

Some people have allergies or sensitivities to foods which are not problematic to most people. This occurs when a person's immune system mistakes a certain food protein for a harmful foreign agent and attacks it. About 2% of adults and 8% of children have a food allergy. The amount of the food substance required to provoke a reaction in a particularly susceptible individual can be quite small. In some instances, traces of food in the air, too minute to be perceived through smell, have been known to provoke lethal reactions in extremely sensitive individuals. Common food allergens are gluten, corn, shellfish (mollusks), peanuts, and soy. Allergens frequently produce symptoms such as diarrhea, rashes, bloating, vomiting, and regurgitation. The digestive complaints usually develop within half an hour of ingesting the allergen.

Rarely, food allergies can lead to a medical emergency, such as anaphylactic shock, hypotension (low blood pressure), and loss of consciousness. An allergen associated with this type of reaction is peanut, although latex products can induce similar reactions. Initial treatment is with epinephrine (adrenaline), often carried by known patients in the form of an Epi-pen or Twinject.

Other Health Issues

Human diet was estimated to cause perhaps around 35% of cancers in a human epidemiological analysis by Richard Doll and Richard Peto in 1981. These cancer may be caused by carcinogens that are present in food naturally or as contaminants. Food contaminated with fungal growth may contain mycotoxins such as aflatoxins which may be found in contaminated corn and peanuts. Other carcinogens identified in food include heterocyclic amines generated in meat when cooked at high temperature, polyaromatic hydrocarbons in charred meat and smoked fish, and nitrosamines generated from nitrites used as food preservatives in cured meat such as bacon.

Anticarcinogens that may help prevent cancer can also be found in many food especially fruits and vegetable. Antioxidants are important groups of compounds that may help remove potentially harmful chemicals. It is however often difficult to identify the specific components in diet that serve to increase or decrease cancer risk since many food, such as beef steak and broccoli, contain low concentrations of both carcinogens and anticarcinogens.

Diet

Other area (Yr 2010) * Africa, sub-Sahara - 2170 kcal/capita/day * N.E. and N. Africa - 3120 kcal/capita/day * South Asia - 2450 kcal/capita/day * East Asia - 3040 kcal/capita/day * Latin America / Caribbean - 2950 kcal/capita/day * Developed countries - 3470 kcal/capita/day

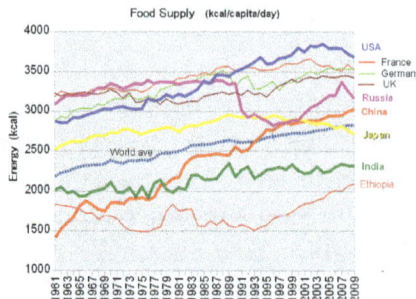

Changes of food supply (by energy)

Cultural and Religious Diets

Dietary habits are the habitual decisions a person or culture makes when choosing what foods to eat. Many cultures hold some food preferences and some food taboos. Dietary choices can also define cultures and play a role in religion. For example, only kosher foods are permitted by Judaism, halal foods by Islam, and in Hinduism beef is restricted. In addition, the dietary choices of different countries or regions have different characteristics. This is highly related to a culture's cuisine.

Diet Deficiencies

Dietary habits play a significant role in the health and mortality of all humans. Imbalances between the consumed fuels and expended energy results in either starvation or excessive reserves of adipose tissue, known as body fat. Poor intake of various vitamins and minerals can lead to diseases that can have far-reaching effects on health. For instance, 30% of the world's population either has, or is at risk for developing, iodine deficiency. It is estimated that at least 3 million children are blind due to vitamin A deficiency. Vitamin C deficiency results in scurvy. Calcium, Vitamin D, and phosphorus are inter-related; the consumption of each may affect the absorption of the others. Kwashiorkor and marasmus are childhood disorders caused by lack of dietary protein.

Moral, Ethical, and Health-conscious Diets

Many individuals limit what foods they eat for reasons of morality, or other habit. For instance, vegetarians choose to forgo food from animal sources to varying degrees. Others choose a healthier diet, avoiding sugars or animal fats and increasing consumption of dietary fiber and antioxidants. Obesity, a serious problem in the western world, leads to higher chances of developing heart disease, diabetes, and many other diseases. More recently, dietary habits have been influenced by the concerns that some people have about possible impacts on health or the environment from genetically modified food. Further concerns about the impact of industrial farming (grains) on animal welfare, human health, and the environment are also having an effect on contemporary human dietary habits. This has led to the emergence of a movement with a preference for organic and local food.

Nutrition and Dietary Problems

Between the extremes of optimal health and death from starvation or malnutrition, there is an array of disease states that can be caused or alleviated by changes in diet. Deficiencies, excesses, and imbalances in diet can produce negative impacts on health, which may lead to various health problems such as scurvy, obesity, or osteoporosis, diabetes, cardiovascular diseases as well as psychological and behavioral problems. The science of nutrition attempts to understand how and why specific dietary aspects influence health.

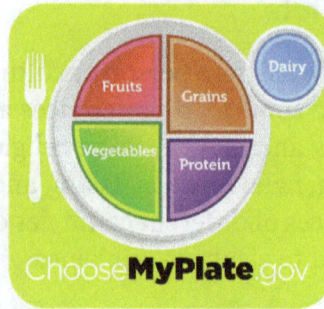

MyPlate replaced MyPyramid as the USDA nutrition guide.

Nutrients in food are grouped into several categories. Macronutrients are fat, protein, and carbohydrates. Micronutrients are the minerals and vitamins. Additionally, food contains water and dietary fiber.

As previously discussed, the body is designed by natural selection to enjoy sweet and fattening foods for evolutionary diets, ideal for hunters and gatherers. Thus, sweet and fattening foods in nature are typically rare and are very pleasurable to eat. In modern times, with advanced technology, enjoyable foods are easily available to consumers. Unfortunately, this promotes obesity in adults and children alike.

Legal Definition

Some countries list a legal definition of food, often referring them with the word *foodstuff*. These countries list food as any item that is to be processed, partially processed, or unprocessed for consumption. The listing of items included as food include any substance intended to be, or reasonably expected to be, ingested by humans. In addition to these foodstuffs, drink, chewing gum, water, or other items processed into said food items are part of the legal definition of food. Items not included in the legal definition of food include animal feed, live animals (unless being prepared for sale in a market), plants prior to harvesting, medicinal products, cosmetics, tobacco and tobacco products, narcotic or psychotropic substances, and residues and contaminants.

Fiber

Fiber or fibre (from the Latin fibra) is a natural or synthetic substance that is significantly longer than it is wide. Fibers are often used in the manufacture of other materials. The strongest engineering materials often incorporate fibers, for example carbon fiber and ultra-high-molecular-weight polyethylene.

Synthetic fibers can often be produced very cheaply and in large amounts compared to natural fibers, but for clothing natural fibers can give some benefits, such as comfort, over their synthetic counterparts.

A bundle of optical fibers

Natural Fibers

Natural fibers develop or occur in the fiber shape, and include those produced by plants, animals, and geological processes. They can be classified according to their origin:

- Vegetable fibers are generally based on arrangements of cellulose, often with lignin: examples include cotton, hemp, jute, flax, ramie, sisal, bagasse, and banana. Plant fibers are employed in the manufacture of paper and textile (cloth), and dietary fiber is an important component of human nutrition.

- Wood fiber, distinguished from vegetable fiber, is from tree sources. Forms include groundwood, lacebark, thermomechanical pulp (TMP), and bleached or unbleached kraft or sulfite pulps. Kraft and sulfite (also called sulphite) refer to the type of pulping process used to remove the lignin bonding the original wood structure, thus freeing the fibers for use in paper and engineered wood products such as fiberboard.

- Animal fibers consist largely of particular proteins. Instances are silkworm silk, spider silk, sinew, catgut, wool, sea silk and hair such as cashmere wool, mohair and angora, fur such as sheepskin, rabbit, mink, fox, beaver, etc.

- Mineral fibers include the asbestos group. Asbestos is the only naturally occurring long mineral fiber. Six minerals have been classified as "asbestos" including chrysotile of the serpentine class and those belonging to the amphibole class: amosite, crocidolite, tremolite, anthophyllite and actinolite. Short, fiber-like minerals include wollastonite and palygorskite.

- Biological fibers also known as fibrous proteins or protein filaments consist largely of biologically relevant and biologically very important proteins, mutations or other genetic defects can lead to severe diseases. Instances are collagen family of proteins, tendon, muscle proteins like actin, cell proteins like microtubules and many others, spider silk, sinew and hair etc.

Man-made Fibers

Man-made fibers or chemical fibers are fibers whose chemical composition, structure, and properties are significantly modified during the manufacturing process. Man-made fibers consist of regenerated fibers and synthetic fibers.

Semi-synthetic Fibers

Semi-synthetic fibers are made from raw materials with naturally long-chain polymer structure and are only modified and partially degraded by chemical processes, in contrast to completely synthetic fibers such as nylon (polyamide) or dacron (polyester), which the chemist synthesizes from low-molecular weight compounds by polymerization (chain-building) reactions. The earliest semi-synthetic fiber is the cellulose regenerated fiber, rayon. Most semi-synthetic fibers are cellulose regenerated fibers.

Cellulose Regenerated Fibers

Cellulose fibers are a subset of man-made fibers, regenerated from natural cellulose. The cellulose comes from various sources: rayon from tree wood fiber, Modal from beech trees, bamboo fiber from bamboo, seacell from seaweed, etc. In the production of these fibers, the cellulose is reduced to a fairly pure form as a viscous mass and formed into fibers by extrusion through spinnerets. Therefore, the manufacturing process leaves few characteristics distinctive of the natural source material in the finished products.

Some examples are:

- rayon
- bamboo fiber
- Lyocell, a brand of rayon
- Modal, using beech trees as input
- diacetate fiber
- triacetate fiber.

Historically, cellulose diacetate and -triacetate were classified under the term rayon, but are now considered distinct materials.

Synthetic Fibers

Synthetic come entirely from synthetic materials such as petrochemicals, unlike those man-made fibers derived from such natural substances as cellulose or protein.

Fiber classification in reinforced plastics falls into two classes: (i) short fibers, also known as discontinuous fibers, with a general aspect ratio (defined as the ratio of fiber length to diameter) between 20 and 60, and (ii) long fibers, also known as continuous fibers, the general aspect ratio is between 200 and 500.

Metallic Fibers

Metallic fibers can be drawn from ductile metals such as copper, gold or silver and extruded or deposited from more brittle ones, such as nickel, aluminum or iron.

Carbon Fiber

Carbon fibers are often based on oxidized and via pyrolysis carbonized polymers like PAN, but the end product is almost pure carbon.

Silicon Carbide Fiber

Silicon carbide fibers, where the basic polymers are not hydrocarbons but polymers, where about 50% of the carbon atoms are replaced by silicon atoms, so-called poly-carbo-silanes. The pyrolysis yields an amorphous silicon carbide, including mostly other elements like oxygen, titanium, or aluminium, but with mechanical properties very similar to those of carbon fibers.

Fiberglass

Fiberglass, made from specific glass, and optical fiber, made from purified natural quartz, are also man-made fibers that come from natural raw materials, silica fiber, made from sodium silicate (water glass) and basalt fiber made from melted basalt.

Mineral Fibers

Mineral fibers can be particularly strong because they are formed with a low number of surface defects, asbestos is a common one.

Polymer Fibers

- Polymer fibers are a subset of man-made fibers, which are based on synthetic chemicals (often from petrochemical sources) rather than arising from natural materials by a purely physical process. These fibers are made from:

    o polyamide nylon

    o PET or PBT polyester

    o phenol-formaldehyde (PF)

- o   polyvinyl chloride fiber (PVC) vinyon

- o   polyolefins (PP and PE) olefin fiber

- o   acrylic polyesters, pure polyester PAN fibers are used to make carbon fiber by roasting them in a low oxygen environment. Traditional acrylic fiber is used more often as a synthetic replacement for wool. Carbon fibers and PF fibers are noted as two resin-based fibers that are not thermoplastic, most others can be melted.

- o   aromatic polyamids (aramids) such as Twaron, Kevlar and Nomex thermally degrade at high temperatures and do not melt. These fibers have strong bonding between polymer chains

- o   polyethylene (PE), eventually with extremely long chains / HMPE (e.g. Dyneema or Spectra).

- o   Elastomers can even be used, e.g. spandex although urethane fibers are starting to replace spandex technology.

- o   polyurethane fiber

- o   Elastolefin

- Coextruded fibers have two distinct polymers forming the fiber, usually as a core-sheath or side-by-side. Coated fibers exist such as nickel-coated to provide static elimination, silver-coated to provide anti-bacterial properties and aluminum-coated to provide RF deflection for radar chaff. Radar chaff is actually a spool of continuous glass tow that has been aluminum coated. An aircraft-mounted high speed cutter chops it up as it spews from a moving aircraft to confuse radar signals.

Microfibers

Microfibers in textiles refer to sub-denier fiber (such as polyester drawn to 0.5 denier). Denier and Dtex are two measurements of fiber yield based on weight and length. If the fiber density is known, you also have a fiber diameter, otherwise it is simpler to measure diameters in micrometers. Microfibers in technical fibers refer to ultra fine fibers (glass or meltblown thermoplastics) often used in filtration. Newer fiber designs include extruding fiber that splits into multiple finer fibers. Most synthetic fibers are round in cross-section, but special designs can be hollow, oval, star-shaped or trilobal. The latter design provides more optically reflective properties. Synthetic textile fibers are often crimped to provide bulk in a woven, non woven or knitted structure. Fiber surfaces can also be dull or bright. Dull surfaces reflect more light while bright tends to transmit light and make the fiber more transparent.

Very short and/or irregular fibers have been called fibrils. Natural cellulose, such as cotton or bleached kraft, show smaller fibrils jutting out and away from the main fiber structure.

Dairy Product

A dairy product or *milk product* is food produced from the milk of mammals, primarily cows, water buffaloes, goats, sheep, yaks, horses, camels, and domestic buffaloes. A facility that processes milk is a dairy or a dairy factory. Dairy products are commonly found in European, Middle Eastern, South Asian, and Central Asian cuisines, but are largely absent from East Asian and Southeast Asian cuisines.

Milk products and production relationships

Types of Dairy Products

A dairy farm

A selection of three common dairy products made by a South African dairy company: a box of full cream, long life milk, a bottle of strawberry drinking yogurt, and a carton of passion fruit yogurt

The milk products of the Water buffaloes (super carabaos, Philippine Carabao Center)

- Milk after optional homogenization, pasteurization, in several grades after standardization of the fat level, and possible addition of the bacteria *Streptococcus lactis* and *Leuconostoc citrovorum*

    o *Crème fraîche*, slightly fermented cream

        - Clotted cream, thick, spoonable cream made by heating milk

        - Single cream, double cream and whipping cream

        - *Smetana*, Central and Eastern European variety of sour cream

    o Cultured milk resembling buttermilk, but uses different yeast and bacterial cultures

    o Kefir, fermented milk drink from the Northern Caucasus

    o *Kumis/Airag*, slightly fermented mares' milk popular in Central Asia

    o Powdered milk (or milk powder), produced by removing the water from (usually skim) milk

        - Whole milk products

        - Buttermilk products

        - Skim milk

        - Whey products

        - High milk-fat and nutritional products (for infant formulas)

        - Cultured and confectionery products

    o Condensed milk, milk which has been concentrated by evaporation, with sugar added for reduced process time and longer life in an opened can

    o *Khoa*, milk which has been completely concentrated by evaporation, used in Indian cuisine including gulab jamun, peda, etc.)

- o Evaporated milk, (less concentrated than condensed) milk without added sugar

- o Ricotta, acidified whey, reduced in volume

- o Infant formula, dried milk powder with specific additives for feeding human infants

- o Baked milk, a variety of boiled milk that has been particularly popular in Russia

- Butter, mostly milk fat, produced by churning cream

    - o Buttermilk, the liquid left over after producing butter from cream, often dried as livestock feed

    - o *Ghee*, clarified butter, by gentle heating of butter and removal of the solid matter

    - o *Smen*, a fermented, clarified butter used in Moroccan cooking

    - o Anhydrous milkfat (clarified butter)

- Cheese, produced by coagulating milk, separating from whey and letting it ripen, generally with bacteria and sometimes also with certain molds

    - o Curds, the soft, curdled part of milk (or skim milk) used to make cheese

    - o *Paneer*

    - o Whey, the liquid drained from curds and used for further processing or as a livestock feed

    - o Cottage cheese

    - o Quark

    - o Cream cheese, produced by the addition of cream to milk and then curdled to form a rich curd or cheese

    - o *Fromage frais*

- Casein are

    - o Caseinates, sodium or calcium salts of casein

    - o Milk protein concentrates and isolates

    - o Whey protein concentrates and isolates, reduced lactose whey

- o Hydrolysates, milk treated with proteolytic enzymes to alter functionality

- o Mineral concentrates, byproduct of demineralizing whey

- Yogurt, milk fermented by *Streptococcus salivarius* ssp. *thermophilus* and *Lactobacillus delbrueckii* ssp. *bulgaricus* sometimes with additional bacteria, such as *Lactobacillus acidophilus*

  - o *Ayran*

  - o *Lassi*, Indian subcontinent

  - o *Leben*

- Clabber, milk naturally fermented to a yogurt-like state

- Gelato, slowly frozen milk and water, lesser fat than ice cream

- Ice cream, slowly frozen cream, milk, flavors and emulsifying additives (dairy ice cream)

  - o Ice milk, low-fat version of ice cream

  - o Frozen custard

  - o Frozen yogurt, yogurt with emulsifiers

- Other

  - o *Viili*

  - o *Kajmak*

  - o *Filmjölk*

  - o *Piimä*

  - o *Vla*

  - o *Dulce de leche*

  - o *Skyr*

  - o Junket, milk solidified with rennet

Health

Dairy products can cause health issues for individuals who have lactose intolerance or a milk allergy.

Additionally dairy products including cheese, ice cream, milk, butter, and yogurt can contribute significant amounts of cholesterol and saturated fat to the diet. Diets high in fat and especially in saturated fat can increase the risk of heart disease and can cause other serious health problems. However, it has been shown that there is no connection between dairy consumption (excluding butter) and cardiovascular disease, even though dairy tends to be higher in saturated fats.

Avoidance

Some groups avoid dairy products for non-health related reasons:

- Religious - Some religions restrict or do not allow for the consumption of dairy products. For example, some scholars of Jainism advocate not consuming any dairy products because dairy is perceived to involve violence against cows. Strict Judaism requires that meat and dairy products not be served at the same meal, served or cooked in the same utensils, or stored together, as prescribed in Deuteronomy 14:21.

- Ethical - Veganism is the avoidance of all animal products, including dairy products, most often due to the ethics regarding how dairy products are produced. The ethical reasons for avoiding dairy include how dairy is produced, how the animals are handled, and the environmental effect of dairy production.

Egg

Eggs are laid by female animals of many different species, including birds, reptiles, amphibians, mammals, and fish, and have been eaten by humans for thousands of years. Bird and reptile eggs consist of a protective eggshell, albumen (egg white), and vitellus (egg yolk), contained within various thin membranes. The most popular choice for egg consumption are chicken eggs. Other popular choices for egg consumption are duck, quail, roe, and caviar.

Eggs of various birds, a reptile, various cartilaginous fish, a cuttlefish and various butterflies and moths.
(Click on image for key)

Egg yolks and whole eggs store significant amounts of protein and choline, and are widely used in cookery. Due to their protein content, the United States Department of Agriculture categorizes eggs as *Meats* within the Food Guide Pyramid. Despite the nutritional value of eggs, there are some potential health issues arising from egg quality, storage, and individual allergies.

Chickens and other egg-laying creatures are widely kept throughout the world, and mass production of chicken eggs is a global industry. In 2009, an estimated 62.1 million metric tons of eggs were produced worldwide from a total laying flock of approximately 6.4 billion hens. There are issues of regional variation in demand and expectation, as well as current debates concerning methods of mass production. The European Union recently banned battery husbandry of chickens.

History

Bird eggs have been valuable foodstuffs since prehistory, in both hunting societies and more recent cultures where birds were domesticated. The chicken was probably domesticated for its eggs from jungle fowl native to tropical and subtropical Southeast Asia and India before 7500 BCE. Chickens were brought to Sumer and Egypt by 1500 BCE, and arrived in Greece around 800 BCE, where the quail had been the primary source of eggs. In Thebes, Egypt, the tomb of Haremhab, built about 1420 BCE, shows a depiction of a man carrying bowls of ostrich eggs and other large eggs, presumably those of the pelican, as offerings. In ancient Rome, eggs were preserved using a number of methods, and meals often started with an egg course. The Romans crushed the shells in their plates to prevent evil spirits from hiding there. In the Middle Ages, eggs were forbidden during Lent because of their richness. The word mayonnaise possibly was derived from *moyeu*, the medieval French word for the yolk, meaning center or hub.

Egg scrambled with acidic fruit juices were popular in France in the 17th century; this may have been the origin of lemon curd.

The dried egg industry developed in the 19th century, before the rise of the frozen egg industry. In 1878, a company in St. Louis, Missouri started to transform egg yolk and white into a light-brown, meal-like substance by using a drying process. The production of dried eggs significantly expanded during World War II, for use by the United States Armed Forces and its allies.

In 1911, the egg carton was invented by Joseph Coyle in Smithers, British Columbia, to solve a dispute about broken eggs between a farmer in Bulkley Valley and the owner of the Aldermere Hotel. Early egg cartons were made of paper.

Varieties

Bird eggs are a common food and one of the most versatile ingredients used in cooking. They are important in many branches of the modern food industry.

Quail eggs (upper left), chicken egg (lower left) and ostrich egg (right)

The most commonly used bird eggs are those from the chicken. Duck and goose eggs, and smaller eggs, such as quail eggs, are occasionally used as a gourmet ingredient in western countries. Eggs are a common everyday food in many parts of Asia, such as China and Thailand, with Asian production providing 59% of the world total in 2013.

The largest bird eggs, from ostriches tend to be used only as special luxury food. Gull eggs are considered a delicacy in England, as well as in some Scandinavian countries, particularly in Norway. In some African countries, guineafowl eggs are commonly seen in marketplaces, especially in the spring of each year. Pheasant eggs and emu eggs are edible, but less widely available. Sometimes they are obtainable from farmers, poulterers, or luxury grocery stores. In many countries, wild birds' eggs are protected by laws which prohibit collecting or selling them, or permit collection only during specific periods of the year.

Global production of chicken eggs (in shell) (2013, in millions of tonnes)	
China	24.8
USA	5.6
India	3.8
Japan	2.5
World	**68.3**
Source: United Nations, Food and Agriculture Organization, FAOSTAT	

Production

In 2013, world production of chicken eggs was 68.3 million tonnes, with China contributing 37% of this total.

Anatomy and Characteristics

The shape of an egg resembles a prolate spheroid with one end larger than the other, with cylindrical symmetry along the long axis.

A raw chicken egg within its membrane, the shell removed by soaking in vinegar

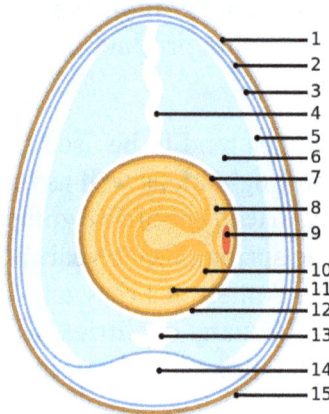

Schematic of a chicken egg:
1. Eggshell, 2. Outer membrane, 3. Inner membrane, 4. Chalaza, 5. Exterior albumen, 6. Middle albumen, 7. Vitelline membrane, 8. Nucleus of pander, 9. Germinal disc (nucleus), 10. Yellow yolk, 11. White yolk 12. Internal albumen, 13. Chalaza, 14. Air cell, 15. Cuticula

An egg is surrounded by a thin, hard shell. Inside, the egg yolk is suspended in the egg white by one or two spiral bands of tissue called the chalazae.

Air Cell

The larger end of the egg contains the air cell that forms when the contents of the egg cool down and contract after it is laid. Chicken eggs are graded according to the size of this air cell, measured during candling. A very fresh egg has a small air cell and receives a grade of AA. As the size of the air cell increases and the quality of the egg decreases, the grade moves from AA to A to B. This provides a way of testing the age of an egg: as the air cell increases in size due to air being drawn through pores in the shell as water is lost, the egg becomes less dense and the larger end of the egg will rise to increasingly shallower depths when the egg is placed in a bowl of water. A very old egg will actually float in the water and should not be eaten.

Shell

Egg shell color is caused by pigment deposition during egg formation in the oviduct and can vary according to species and breed, from the more common white or brown to pink or speckled blue-green. In general, chicken breeds with white ear lobes lay white eggs, whereas chickens with red ear lobes lay brown eggs. Although there is no significant link between shell color and nutritional value, there is often a cultural preference for one color over another.

Membrane

The membrane is a clear film lining the eggshell, visible when one peels a boiled egg. Eggshell membrane is primarily composed of fibrous proteins such as collagen type I.

White

White is the common name for the clear liquid (also called the albumen or the glair/glaire) contained within an egg. In chickens it is formed from the layers of secretions of the anterior section of the hen's oviduct during the passage of the egg. It forms around either fertilized or unfertilized yolks. The primary natural purpose of egg white is to protect the yolk and provide additional nutrition for the growth of the embryo.

Egg white consists primarily of about 90% water into which is dissolved 10% proteins (including albumins, mucoproteins, and globulins). Unlike the yolk, which is high in lipids (fats), egg white contains almost no fat and the carbohydrate content is less than 1%. Egg white has many uses in food, and many others, including the preparation of vaccines such as those for influenza.

Yolk

The yolk in a newly laid egg is round and firm. As the yolk ages, it absorbs water from the albumen, which increases its size and causes it to stretch and weaken the vitelline membrane (the clear casing enclosing the yolk). The resulting effect is a flattened and enlarged yolk shape.

Yolk color is dependent on the diet of the hen; if the diet contains yellow/orange plant pigments known as xanthophylls, then they are deposited in the yolk, coloring it. Lutein is the most abundant pigment in egg yolk. A colorless diet can produce an almost colorless yolk. Yolk color is, for example, enhanced if the diet includes products such as yellow corn and marigold petals. In the US, the use of artificial color additives is forbidden.

Abnormalities

Shell-less or thin-shelled eggs can be caused by egg drop syndrome.

Culinary Properties

Types of Dishes

Chicken eggs are widely used in many types of dishes, both sweet and savory, including many baked goods. Some of the most common preparation methods include scrambled, fried, hard-boiled, soft-boiled, omelettes and pickled. They can also be eaten raw, though this is not recommended for people who may be especially susceptible to salmonellosis, such as the elderly, the infirm, or pregnant women. In addition, the protein in raw eggs is only 51% bioavailable, whereas that of a cooked egg is nearer 91% bio-available, meaning the protein of cooked eggs is nearly twice as absorbable as the protein from raw eggs.

A fried chicken egg, "sunny side up"

As an ingredient, egg yolks are an important emulsifier in the kitchen, and are also used as a thickener in custards.

Soft-boiled quail eggs, with potato galettes

The albumen, or egg white, contains protein, but little or no fat, and can be used in cooking separately from the yolk. The proteins in egg white allow it to form foams and aerated dishes. Egg whites may be aerated or whipped to a light, fluffy consistency, and are often used in desserts such as meringues and mousse.

Ground egg shells are sometimes used as a food additive to deliver calcium. Every part of an egg is edible, although the eggshell is generally discarded. Some recipes call for immature or unlaid eggs, which are harvested after the hen is slaughtered or cooked while still inside the chicken.

Cooking

Eggs contain multiple proteins which gel at different temperatures within the yolk and the white, and the temperature determines the gelling time. Egg yolk begins to gclify, or solidify, when it reaches temperatures between about 60 and 70 °C (140 and 158 °F). Egg white gels at slightly higher temperatures, about 60 to 80 °C (140 to 176 °F)- the white contains ovalbumin that sets at the highest temperature. However, in practice, in many cooking processes the white gels first because it is exposed to higher temperatures for longer.

A few tips for baking with eggs and cakes.

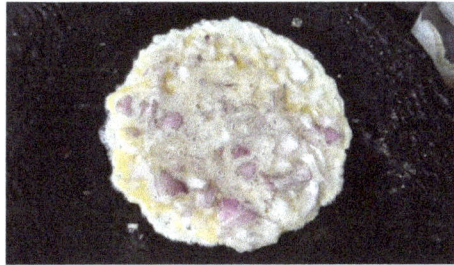

half boiled egg dish

Salmonella is killed instantly at 71 °C (160 °F), but is also killed from 54.5 °C (130.1 °F) if held there for sufficiently long time periods. To avoid the issue of salmonella, eggs can be pasteurized in-shell at 57 °C (135 °F) for an hour and 15 minutes. Although the white is slightly milkier, the eggs can be used in normal ways. Whipping for meringue takes significantly longer, but the final volume is virtually the same.

If a boiled egg is overcooked, a greenish ring sometimes appears around egg yolk due to the iron and sulfur compounds in the egg. It can also occur with an abundance of iron in the cooking water. The green ring does not affect the egg's taste; overcooking, however, harms the quality of the protein. Chilling the egg for a few minutes in cold water until it is completely cooled may prevent the greenish ring from forming on the surface of the yolk.

Flavor Variations

Although the age of the egg and the conditions of its storage have a greater influence, the bird's diet does affect the flavor of the egg. For example, when a brown-egg chicken breed eats rapeseed or soy meals, its intestinal microbes metabolize them into fishy-smelling triethylamine, which ends up in the egg. The unpredictable diet of free-range hens will produce unpredictable eggs. Duck eggs tend to have a flavor distinct from, but still resembling chicken eggs.

A batch of tea eggs with shell intact soaking in a brew of spices and tea

Eggs can also be soaked in mixtures to absorb flavor. Tea eggs are steeped in a brew from a mixture of various spices, soy sauce, and black tea leaves to give flavor.

Storage

Careful storage of edible eggs is extremely important, as an improperly handled egg can contain elevated levels of *Salmonella* bacteria that can cause severe food poisoning. In the US, eggs are washed, and this cleans the shell, but erodes the cuticle. The USDA thus recommends refrigerating eggs to prevent the growth of *Salmonella*.

Refrigeration also preserves the taste and texture. However, intact eggs can be left unrefrigerated for several months without spoiling. In Europe, eggs are not usually washed, and the shells are dirtier, however the cuticle is undamaged, and they do not require refrigeration. In the UK in particular, hens are immunized against salmonella, and the eggs are generally safe for 21 days.

Preservation

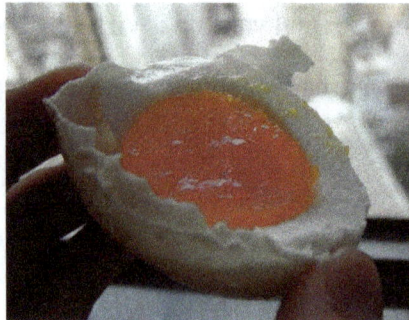

Salted duck egg

The simplest method to preserve an egg is to treat it with salt. Salt draws water out of bacteria and molds, which prevents their growth. The Chinese salted duck egg is made by immersing duck eggs in brine, or coating them individually with a paste of salt and mud or clay. The eggs stop absorbing salt after about a month, having reached osmotic equilibrium. Their yolks take on an orange-red color and solidify, but the white remains somewhat liquid. They are often boiled before consumption, and are often served with rice congee.

Pickled egg, colored with beetroot juice

Another method is to make pickled eggs, by boiling them first and immersing them in a mixture of vinegar, salt and spices, such as ginger or allspice. Frequently, beetroot juice is added to impart a red color to the eggs. If the eggs are immersed in it for a few hours, the distinct red, white, and yellow colors can be seen when the eggs are sliced. If marinated for several days or more, the red color will reach the yolk. If the eggs are marinated in the mixture for several weeks or more, the vinegar will dissolve much of the shell's calcium carbonate and penetrate the egg, making it acidic enough to inhibit the growth of bacteria and molds. Pickled eggs made this way will generally keep for a year or more without refrigeration.

Century egg

A century egg or hundred-year-old egg is preserved by coating an egg in a mixture of clay, wood ash, salt, lime, and rice straw for several weeks to several months, depending on the method of processing. After the process is completed, the yolk becomes a dark green, cream-like substance with a strong odor of sulfur and ammonia, while the white becomes a dark brown, transparent jelly with a comparatively mild, distinct flavor. The transforming agent in a century egg is its alkaline material, which gradually raises the pH of the egg from around 9 to 12 or more. This chemical process breaks down some of the complex, flavorless proteins and fats of the yolk into simpler, flavorful ones, which in some way may be thought of as an "inorganic" version of fermentation.

Cooking Substitutes

For those who do not consume eggs, alternatives used in baking include other rising agents or binding materials, such as ground flax seeds or potato starch flour. Tofu can also act as a partial binding agent, since it is high in lecithin due to its soy content. Applesauce can be used, as well as arrowroot and banana. Extracted soybean lecithin, in turn, is often used in packaged foods as an inexpensive substitute for egg-derived lecithin. Chickpea brine, also known as aquafaba, can replace egg whites in desserts like meringues and mousses.

Other egg substitutes are made from just the white of the egg for those who worry about the high cholesterol and fat content in eggs. These products usually have added vitamins and minerals, as well as vegetable-based emulsifiers and thickeners such as xanthan gum or guar gum. These allow the product to maintain the nutrition and several culinary properties of real eggs, making possible foods such as Hollandaise sauce, custard, mayonnaise, and most baked goods with these substitutes.

Nutritional Value

Chicken eggs, the most commonly eaten eggs, provide 155 calories (kcal) of food energy and 12.6 g of protein in a 100 gram serving.

Eggs (boiled) supply several vitamins and minerals as significant amounts of the Daily Value (DV), including vitamin A (19% DV), riboflavin (42% DV), pantothenic acid (28% DV), vitamin B_{12} (46% DV), choline (60% DV), phosphorus (25% DV), zinc (11% DV) and vitamin D (15% DV) (table per 100 gram serving of a hard-boiled egg).

A yolk contains more than two-thirds of the recommended daily intake of 300 mg of cholesterol (table).

The diet of laying hens can affect the nutritional quality of eggs. For instance, chicken eggs that are especially high in omega-3 fatty acids are produced by feeding hens a diet containing polyunsaturated fats from sources like fish oil, chia seeds or flaxseeds. Pasture-raised free-range hens, which forage for their own food, also produce eggs that are relatively enriched in omega-3 fatty acids compared to cage-raised chickens.

A 2010 USDA study determined there were no significant differences of macronutrients in various chicken eggs.

Cooked eggs are easier to digest, as well as having a lower risk of salmonellosis.

Health Effects

Cholesterol And Fat

More than half the calories found in eggs come from the fat in the yolk; 50 grams of chicken egg (the contents of an egg just large enough to be classified as "large" in the US, but "medium" in Europe) contains approximately 5 grams of fat. People on a low-cholesterol diet may need to reduce egg consumption; however, only 27% of the fat in egg is saturated fat (palmitic, stearic and myristic acids). The egg white consists primarily of water (87%) and protein (13%) and contains no cholesterol and little, if any, fat.

There is debate over whether egg yolk presents a health risk. Some research suggests dietary cholesterol increases the ratio of total to HDL cholesterol and, therefore, adversely affects the body's cholesterol profile; whereas other studies show that moderate consumption of eggs, up to one a day, does not appear to increase heart disease risk in healthy individuals. Harold McGee argues that the cholesterol in the yolk is not what causes a problem, because fat (in particular, saturated) is much more likely to raise cholesterol levels than the actual consumption of cholesterol.

Type 2 Diabetes

Studies have shown conflicting results about a possible connection between egg consumption and type two diabetes. A 1999 prospective study of over 117,000 people by the Harvard School of Public Health concluded, in part, that "The apparent increased risk of CHD associated with higher egg consumption among diabetic participants warrants further research." A 2008 study by the Physicians' Health Study I (1982–2007) and the Women's Health Study (1992–2007) determined the "data suggest that high levels of egg consumption (daily) are associated with an increased risk of type 2 diabetes." However, a study published in 2010 found no link between egg consumption and type 2 diabetes. A meta-analysis from 2013 found that eating 4 eggs per week increased the risk of diabetes by 29%. Another meta-analysis from 2013 also supported the idea that egg consumption may lead to an increased incidence of type two diabetes.

Cardiovascular Risk

Eggs are one of the largest sources of phosphatidylcholine (lecithin) in the human diet. A study published in the scientific journal Nature showed that dietary phosphatidylcholine is digested by bacteria in the gut and eventually converted into the compound TMAO, a compound linked with increased heart disease.

The 1999 Harvard School of Public Health study of 37,851 men and 80,082 women concluded that its "findings suggest that consumption of up to 1 egg per day is unlikely to have substantial overall impact on the risk of CHD or stroke among healthy men and women." In a study of 4,000 people, scientists found that eating eggs increased blood levels of a metabolite promoting atherosclerosis, TMAO, and that this in turn caused significantly higher risk of heart attack and stroke after three years of follow-up.

A 2007 study of nearly 10,000 adults demonstrated no correlation between moderate (six per week) egg consumption and cardiovascular disease or strokes, except in the subpopulation of diabetic patients who presented an increased risk of coronary artery disease. One potential alternative explanation for the null finding is that background dietary cholesterol may be so high in the usual Western diet that adding somewhat more has little further effect on blood cholesterol. Other research supports the idea that a high egg intake increases cardiovascular risk in diabetic patients. A 2009 prospective cohort study of over 21,000 individuals suggests that "egg consumption up to 6/week has no major effect on the risk of cardiovascular disease and mortality and that consumption of 7+/week is associated with a modest increased risk of total mortality" in males, whereas among males with diabetes, "any egg consumption is associated with an increased risk of all-cause mortality and there was suggestive evidence for an increased risk of myocardial infarction and stroke". A 2013 meta-analysis found no association between egg consumption and heart disease or stroke. A 2013 systematic review and meta-analysis found no association between egg consumption and cardiovascular disease or cardiovascular disease mortality, but did find egg consumption more than once daily increased cardiovascular disease risk 1.69-fold in those with type 2 diabetes mellitus compared to type 2 diabetics who ate less than 1 egg per week. Another 2013 meta-analysis found that eating 4 eggs per week increased the risk of cardiovascular disease by 6%.

Contamination

A health issue associated with eggs is contamination by pathogenic bacteria, such as *Salmonella enteritidis*. Contamination of eggs exiting a female bird via the cloaca may also occur with other members of the *Salmonella* genus, so care must be taken to prevent the egg shell from becoming contaminated with fecal matter. In commercial practice in the US, eggs are quickly washed with a sanitizing solution within minutes of being laid. The risk of infection from raw or undercooked eggs is dependent in part upon the sanitary conditions under which the hens are kept.

Egg cleaning on a farm in Norway

Health experts advise people to refrigerate washed eggs, use them within two weeks, cook them thoroughly, and never consume raw eggs. As with meat, containers and surfaces that have been used to process raw eggs should not come in contact with ready-to-eat food.

A study by the U.S. Department of Agriculture in 2002 (Risk Analysis April 2002 22(2):203-18) suggests the problem is not as prevalent as once thought. It showed that of the 69 billion eggs produced annually, only 2.3 million are contaminated with *Salmonella*—equivalent to just one in every 30,000 eggs—thus showing *Salmonella* infection is quite rarely induced by eggs. However, this has not been the case in other countries, where *Salmonella enteritidis* and *Salmonella typhimurium* infections due to egg consumptions are major concerns. Egg shells act as hermetic seals that guard against bacteria entering, but this seal can be broken through improper handling or if laid by unhealthy chickens. Most forms of contamination enter through such weaknesses in the shell. In the UK, the British Egg Industry Council award the lions stamp to eggs that, among other things, come from hens that have been vaccinated against *Salmonella*.

Food Allergy

One of the most common food allergies in infants is eggs. Infants usually have the opportunity to grow out of this allergy during childhood, if exposure is minimized. Allergic reactions against egg white are more common than reactions against egg yolks.

In addition to true allergic reactions, some people experience a food intolerance to egg whites.

Food labeling practices in most developed countries now include eggs, egg products and the processing of foods on equipment that also process foods containing eggs in a special allergen alert section of the ingredients on the labels.

Farming Issues

Most commercially farmed chicken eggs intended for human consumption are unfertilized, since the laying hens are kept without roosters. Fertile eggs can be eaten, with a little nutritional difference to the unfertilized. Fertile eggs will not contain a developed embryo, as refrigeration temperatures inhibit cellular growth for an extended time. Sometimes an embryo is allowed to develop but eaten before hatching as with balut.

Eggs for sale at a grocery store

White and brown eggs in an egg crate.

Grading by Quality and Size

The US Department of Agriculture grades eggs by the interior quality of the egg and the appearance and condition of the egg shell. Eggs of any quality grade may differ in weight (size).

- U.S. Grade AA

    o Eggs have whites that are thick and firm; yolks that are high, round, and practically free from defects; and clean, unbroken shells.

    o Grade AA and Grade A eggs are best for frying and poaching, where appearance is important.

- U.S. Grade A

    o Eggs have characteristics of Grade AA eggs except the whites are "reasonably" firm.

    o This is the quality most often sold in stores.

- U.S. Grade B

    o Eggs have whites that may be thinner and yolks that may be wider and flatter than eggs of higher grades. The shells must be unbroken but may show slight stains.

    o This quality is seldom found in retail stores because they are usually used to make liquid, frozen, and dried egg products, as well as other egg-containing products.

In Australia and the European Union, eggs are graded by the hen farming method, free range, battery caged, etc.

Chicken eggs are also graded by size for the purpose of sales. Some maxi eggs can have double-yolk and some farms separate out double-yolk eggs for special sale.

Comparison of an egg and a maxi egg with a double-yolk - Closed (1/2)

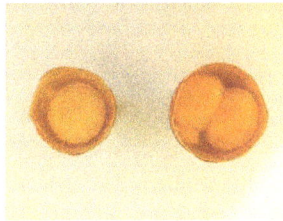

Comparison of an egg and a maxi egg with a double-yolk - Opened (2/2)

Double-yolk egg - Opened

Washing and Refrigeration

In North America, the legislation requires eggs to be washed and refrigerated before being sold to consumers. This is to remove natural farm contaminants present in the cleanest farms and to prevent the growth of bacteria. In Europe legislation requires the opposite. Washing removes the natural protective cuticle on the egg and refrigeration causes condensation which may promote bacteria growth.

Color of Eggshell

Although egg color is a largely cosmetic issue, with no effect on egg quality or taste, it is a major issue in production due to regional and national preferences for specific colors, and the results of such preferences on demand. For example, in most regions of the United States, chicken eggs are generally white. In some parts of the northeast of that country, particularly New England, where a television jingle for years proclaimed "brown eggs are local eggs, and local eggs are fresh!", brown eggs are more common. Local chicken breeds, including the Rhode Island Red, lay brown eggs. Brown eggs are also preferred in Costa Rica, Ireland, France, and the United Kingdom. In Brazil and Poland, white chicken eggs are generally regarded as industrial, and brown or red-dish ones are preferred. Small farms and smallholdings, particularly in economically advanced nations, may sell eggs of widely varying colors and sizes, with combinations of white, brown, speckled (red), green, and blue (as laid by certain breeds, including araucanas, heritage skyline and cream leg bar) eggs in the same box or carton, while the supermarkets at the same time sell mostly eggs from the larger producers, of the color preferred in that nation or region.

White, speckled (red), and brown chicken eggs

Very dark brown eggs of Marans, a French breed of chicken

These cultural trends have been observed for many years. *The New York Times* reported during the Second World War that housewives in Boston preferred brown eggs and those in New York preferred white eggs. In February 1976, the New Scientist magazine, in discussing issues of chicken egg color, stated "Housewives are particularly fussy about the colour of their eggs, preferring even to pay more for brown eggs although white eggs are just as good". As a result of these trends, brown eggs are usually more expensive to purchase in regions where white eggs are considered 'normal', due to lower production. In France and the United Kingdom, it is very difficult to buy white eggs, with most supermarkets supplying only the more popular brown eggs. By direct contrast, in Egypt it is very hard to source brown eggs, as demand is almost entirely for white ones, with the country's largest supplier describing white eggs as "table eggs" and packaging brown eggs for export.

Research conducted by a France institute in the 1970s demonstrated blue chicken eggs from the Chilean araucana fowl can be stronger and more resilient to breakage.

Research at Nihon University, Japan in 1990 revealed a number of different issues were important to Japanese housewives when deciding which eggs to buy; however, color was a distinct factor, with most Japanese housewives preferring the white color.

Egg producers carefully consider cultural issues, as well as commercial ones, when selecting the breed or breeds of chicken used for production, as egg color varies between breeds. Among producers and breeders, brown eggs are often referred to as "tinted", while the speckled eggs preferred by some consumers are often referred to as being "red" in color.

Living Conditions of Birds

Commercial factory farming operations often involve raising the hens in small, crowded cages, preventing the chickens from engaging in natural behaviors, such as wing-flapping, dust-bathing, scratching, pecking, perching, and nest-building. Such restrictions can lead to pacing and escape behavior.

Laying hens in battery cages

Many hens confined to battery cages, and some raised in cage-free conditions, are debeaked to prevent them from harming each other and cannibalism. According to critics of the practice, this can cause hens severe pain to the point where some may refuse to eat and starve to death. Some hens may be force molted to increase egg quality and production level after the molting. Molting can be induced by extended feed withdrawal, water withdrawal or controlled lighting programs.

Laying hens are often slaughtered between 100 and 130 weeks of age, when their egg productivity starts to decline. Due to modern selective breeding, laying hen strains differ from meat production strains. As male birds of the laying strain do not lay eggs and are not suitable for meat production, they are generally killed soon after they hatch.

Free-range eggs are considered by some advocates to be an acceptable substitute to factory-farmed eggs. Free-range laying hens are given outdoor access instead of being contained in crowded cages. Questions on the actual living conditions of free-range hens have been raised in the United States of America, as there is no legal definition or regulations for eggs labeled as free-range in that country.

In the United States, increased public concern for animal welfare has pushed various egg producers to promote eggs under a variety of standards. The most widespread standard in use is determined by United Egg Producers through their voluntary program of certification. The United Egg Producers program includes guidelines regarding housing, food, water, air, living space, beak trimming, molting, handling, and transportation, however, opponents such as The Humane Society have alleged UEP Certification is misleading and allows a significant amount of unchecked animal cruelty. Other standards include "Cage Free", "Natural", "Certified Humane", and "Certified Organic". Of these standards, "Certified Humane", which carries requirements for stocking density and cage-free keeping and so on, and "Certified Organic", which requires hens to have outdoor access and be fed only organic vegetarian feed and so on, are the most stringent.

Effective 1 January 2012, the European Union banned conventional battery cages for egg-laying hens, as outlined in EU Directive 1999/74/EC. The EU permits the use of "enriched" furnished cages that must meet certain space and amenity requirements. Egg producers in many member states have objected to the new quality standards while in some countries even furnished cages and family cages are subject to be banned as well. The production standard of the eggs is visible on the mandatory egg marking where the EU egg code begins with 3 for caged chicken to 1 for free-range eggs and 0 for organic egg production.

Killing of Male Chicks

In battery cage and free-range egg production, unwanted male chicks are killed at birth during the process of securing a further generation of egg-laying hens.

Cultural Significance

A popular Easter tradition in some parts of the world is a decoration of hard-boiled eggs (usually by dying, but often by spray-painting). Adults often hide the eggs for children to find, an activity known as an Easter egg hunt. A similar tradition of egg painting exists in areas of the world influenced by the culture of Persia. Before the spring equinox in the Persian New Year tradition (called *Norouz*), each family member decorates a hard-boiled egg and sets them together in a bowl.

Hanácké kraslice, Easter eggs from the Haná region, the Czech Republic

The tradition of a dancing egg is held during the feast of Corpus Christi in Barcelona and other Catalan cities since the 16th century. It consists of an emptied egg, positioned over the water jet from a fountain, which starts turning without falling.

Although a food item, eggs are sometimes thrown at houses, cars, or people. This act, known commonly as "egging" in the various English-speaking countries, is a minor form of vandalism and, therefore, usually a criminal offense and is capable of damaging property (egg whites can degrade certain types of vehicle paint) as well as causing serious eye injury. On Halloween, for example, trick or treaters have been known to throw eggs (and sometimes flour) at property or people from whom they received nothing. Eggs are also often thrown in protests, as they are inexpensive and nonlethal, yet very messy when broken.

Fur

Fur is used in reference to the hair of animals, usually mammals, particularly those with extensive body hair coverage that is generally soft and thick, as opposed to the stiffer bristles on most pigs. The term *pelage* – first known use in English c. 1828 – (French, from Middle French, from *poil* hair, from Old French *peilss*, from Latin *pilus*,) is sometimes used to refer to the body hair of an animal as a complete coat. *Fur* is also used to refer to animal pelts which have been processed into leather with the hair still attached. The words *fur* or *furry* are also used, more casually, to refer to hair-like growths or formations, particularly when the subject being referred to exhibits a dense coat of fine, soft "hairs." If layered, rather than grown as a single coat, it may consist of short down hairs, long guard hairs, and, in some cases, medium awn hairs. Mammals with reduced amounts of fur are often called "naked", as with the naked mole-rat, or "hairless", as with hairless dogs.

Opossum fur

An animal with commercially valuable fur is known within the fur industry as a furbearer. The use of fur as clothing and/or decoration is considered controversial by some people: most animal welfare advocates object to the trapping and killing of wildlife, and to the confinement and killing of animals on fur farms.

Composition

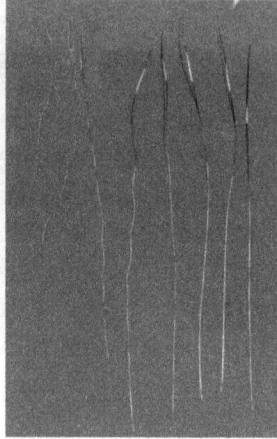

Down, awn and guard hairs of a domestic tabby cat

Fur usually consists of two main layers:

- Down hair (known also as undercoat or ground hair) — the bottom layer consisting of wool hairs, usually wavy or curly without straight portions or sharp points; down hairs tend to be shorter, flat, curly, and more numerous than the top layer. Its principal function is thermoregulation; it maintains a layer of dry air next to the skin and repels water, thus providing thermal insulation.

- Guard hair — the top layer consisting of longer, generally coarser, nearly straight shafts of hair that protrude through the down hair layer. The distal ends of the guard hairs provide the externally visible layer of the coat of most mammals with well-developed fur. This layer of the coat displays the most marked pigmentation and gloss, including coat patterns adapted to display or camouflage. It is also adapted to shedding water and blocking sunlight, protecting the undercoat and skin from external factors such as rain and ultraviolet radiation. Many animals, such as domestic cats, erect their guard hairs as part of their threat display when agitated.

Mammals with well-developed down and guard hairs also usually have large numbers of awn hairs. These begin their growth much as guard hairs do, but change their mode of growth, usually when less than half the length of the hair has emerged. This portion of the hair is called awn. The rest of the growth is thin and wavy, much like down hair. In many species of mammals, the awn hairs comprise the bulk of the visible coat. The proximal part of the awn hair shares the function of the down hairs, whereas the distal part aids the water-shedding function of the guard hairs, though their thin basal portion prevents their being erected like true guard hairs.

The modern fur arrangement is known to have occurred as far back as docodonts, haramiyidans and eutriconodonts, with *Castorocauda*, *Megaconus* and *Spinolestes* preserving compound follicles with guard hair and underfur.

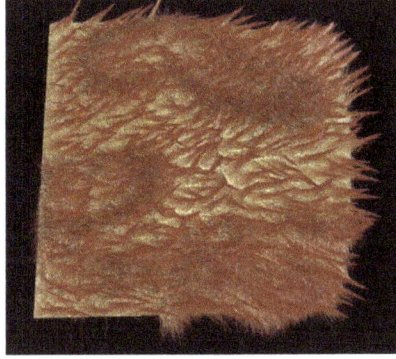
Computer-generated wet fur

Mammals without Fur

Hair is one of the defining characteristics of mammals, however, several species or breeds have considerably reduced amounts of fur. These are often called "naked" or "hairless".

Natural Selection

Some mammals naturally have reduced amounts of fur. Some semiaquatic or aquatic mammals such as cetaceans, pinnipeds and hippopotamuses have evolved hairlessness, presumably to reduce resistance through water. The naked mole-rat has evolved hairlessness, perhaps as an adaptation to their subterranean life-style. Two of the largest extant mammals, the elephant and the rhinoceros, are largely hairless. The hairless bat is mostly hairless but does have short bristly hairs around its neck, on its front toes, and around the throat sac, along with fine hairs on the head and tail membrane. Most hairless animals cannot go in the sun for long periods of time, or stay in the cold for too long.

Humans are the only primate species that have undergone significant hair loss. The hairlessness of humans compared to related species may be due to loss of functionality in the pseudogene KRTHAP1 (which helps produce keratin) in the human lineage about 240,000 years ago. Mutations in the gene HR can lead to complete hair loss, though this is not typical in humans.

Sheep have not become hairless, however, their pelage is usually referred to as "wool" rather than fur.

Artificial Selection

Humans have artificially selected some domesticated mammalian species to have breeds that are hairless. There are several breeds of hairless cats, perhaps the most commonly known being the Sphynx cat. Similarly, there are several breeds of hairless dogs. Other examples of artificially selected hairless animals include the hairless guinea-pig, nude mouse, and the hairless rat.

Use in Clothing

In clothing, fur is usually leather with the hair retained for its aesthetic and insulating properties. Fur has long served as a source of clothing for hominoids including the Neanderthal. Animal furs used in garments and trim may be dyed bright colors or to mimic exotic animal patterns, or shorn down to imitate the feel of a soft velvet fabric. The term "a fur" is often used to refer to a fur coat, wrap, or shawl.

Carl Ben Eielson, US Pilot and Arctic explorator wearing a seal fur coat

Usual animal sources for fur clothing and fur trimmed accessories include fox, rabbit, mink, beavers, ermine, otters, sable, seals, coyotes, chinchilla, raccoon, and possum. The import and sale of seal products was banned in the U.S. in 1972 over conservation concerns about Canadian seals. The import and sale is still banned even though the Marine Animal Response Society estimates the harp seal population is thriving at approximately 8 million. The import, export and sales of domesticated cat and dog fur were also banned in the U.S. under the Dog and Cat Protection Act of 2000.

The manufacturing of fur clothing involves obtaining animal pelts where the hair is left on the animal's processed skin. In contrast, making leather involves removing the hair from the hide or pelt and using only the skin. The use of wool involves shearing the animal's fleece from the living animal, so that the wool can be regrown but sheepskin shearling is made by retaining the fleece to the leather and shearing it. Shearling is used for boots, jackets and coats and is probably the most common type of skin worn.

Fur is also used to make felt. A common felt is made from beaver fur and is used in high-end cowboy hats.

Controversy

Most animal rights activists are opposed to the trapping and killing of wildlife, and the confinement and killing of animals on fur farms. According to Humane Society International, over 8 million animals are trapped yearly for fur, while more than 30 million were raised in fur farms.

Red fox furs

According to Statistics Canada, 2.6 million fur-bearing animals raised on farms were killed in 2010. Another 700,000 were killed for fur by traps.

Wool

Wool is the textile fiber obtained from sheep and certain other animals, including cashmere from goats, mohair from goats, qiviut from muskoxen, angora from rabbits, and other types of wool from camelids.

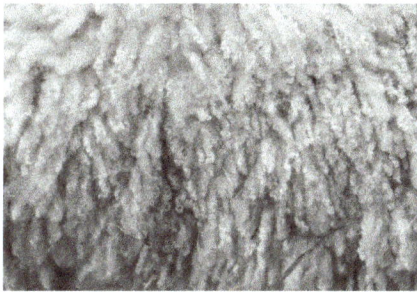
Wool just before processing

Sheep before shearing

Merino sheep

Wool has several qualities that distinguish it from hair or fur: it is crimped, it is elastic, and it grows in staples (clusters).

Characteristics

Wool's scaling and crimp make it easier to spin the fleece by helping the individual fibers attach to each other, so they stay together. Because of the crimp, wool fabrics have greater bulk than other textiles, and they hold air, which causes the fabric to retain heat. Wool has a high specific heat coefficient, so it impedes heat transfer in general. This effect has benefited desert peoples, as Bedouins and Tuaregs use wool clothes for insulation.

Champion hogget fleece, Walcha Show

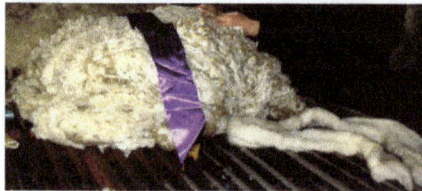
Fleece of fine New Zealand Merino wool and combed wool top on a wool table

Felting of wool occurs upon hammering or other mechanical agitation as the microscopic barbs on the surface of wool fibers hook together.

The amount of crimp corresponds to the fineness of the wool fibers. A fine wool like Merino may have up to 100 crimps per inch, while the coarser wools like karakul may have as few as one or two. In contrast, hair has little if any scale and no crimp, and little ability to bind into yarn. On sheep, the hair part of the fleece is called kemp. The relative amounts of kemp to wool vary from breed to breed and make some fleeces more desirable for spinning, felting, or carding into batts for quilts or other insulating products, including the famous tweed cloth of Scotland.

Wool fibers readily absorb moisture, but are not hollow. Wool can absorb almost one-third of its own weight in water. Wool absorbs sound like many other fabrics. It is generally a creamy white color, although some breeds of sheep produce natural colors, such as black, brown, silver, and random mixes.

Wool ignites at a higher temperature than cotton and some synthetic fibers. It has a lower rate of flame spread, a lower rate of heat release, a lower heat of combustion, and does not melt or drip; it forms a char which is insulating and self-extinguishing, and it contributes less to toxic gases and smoke than other flooring products when used in carpets. Wool carpets are specified for high safety environments, such as trains and aircraft. Wool is usually specified for garments for firefighters, soldiers, and others in occupations where they are exposed to the likelihood of fire.

Wool is considered by the medical profession to be allergenic.

Processing

Shearing

Sheep shearing is the process by which the woolen fleece of a sheep is cut off. After shearing, the wool is separated into four main categories: fleece (which makes up the vast bulk), broken, bellies, and locks. The quality of fleeces is determined by a technique known as wool classing, whereby a qualified person called a wool classer groups wools of similar gradings together to maximize the return for the farmer or sheep owner. In Australia before being auctioned, all Merino fleece wool is objectively measured for micron, yield (including the amount of vegetable matter), staple length, staple strength, and sometimes color and comfort factor. The sheep is given a dip in antiseptic solution after shearing, so as to cure the wounds caused during shearing.

Fine Merino shearing Lismore, Victoria

Scouring

Wool straight off a sheep, known as "greasy wool" or "wool in the grease", contains a high level of valuable lanolin, as well as the sheep's dead skin and sweat residue, and generally also contains pesticides and vegetable matter from the animal's environment. Before the wool can be used for commercial purposes, it must be scoured, a process of cleaning the greasy wool. Scouring may be as simple as a bath in warm water or as complicated as an industrial process using detergent and alkali in specialized equipment. In north west England, special potash pits were constructed to produce potash used in the manufacture of a soft soap for scouring locally produced white wool.

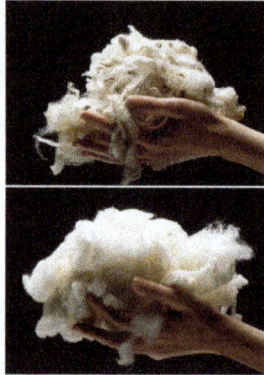

Wool before and after scouring

In commercial wool, vegetable matter is often removed by chemical carbonization. In less-processed wools, vegetable matter may be removed by hand and some of the lanolin left intact through the use of gentler detergents. This semigrease wool can be worked into yarn and knitted into particularly water-resistant mittens or sweaters, such as those of the Aran Island fishermen. Lanolin removed from wool is widely used in cosmetic products such as hand creams.

Quality

The quality of wool is determined by its fiber diameter, crimp, yield, color, and staple strength. Fiber diameter is the single most important wool characteristic determining quality and price.

Various types and natural colors of wool, and a picture made from wool

Merino wool is typically 3–5 inches in length and is very fine (between 12 and 24 microns). The finest and most valuable wool comes from Merino hoggets. Wool taken from sheep produced for meat is typically more coarse, and has fibers 1.5 to 6 in (38 to 152 mm) in length. Damage or breaks in the wool can occur if the sheep is stressed while it is growing its fleece, resulting in a thin spot where the fleece is likely to break.

Wool is also separated into grades based on the measurement of the wool's diameter in microns and also its style. These grades may vary depending on the breed or purpose of the wool. For example:

- < 15.5: Ultrafine Merino

- 15.6 – 18.5: Superfine Merino

- 18.6 – 20: Fine Merino

- 20.1 – 23: Medium Merino

- > 23: Strong Merino

- Comeback: 21–26 microns, white, 90–180 mm long

- Fine crossbred: 27–31 microns, Corriedales, etc.

- Medium crossbred: 32–35 microns

- Downs: 23–34 microns, typically lacks luster and brightness. Examples, Aussiedown, Dorset Horn, Suffolk, etc.

- Coarse crossbred: >36 microns

- Carpet wools: 35–45 microns

Any wool finer than 25 microns can be used for garments, while coarser grades are used for outerwear or rugs. The finer the wool, the softer it is, while coarser grades are more durable and less prone to pilling.

The finest Australian and New Zealand Merino wools are known as 1PP, which is the industry benchmark of excellence for Merino wool 16.9 microns and finer. This style represents the top level of fineness, character, color, and style as determined on the basis of a series of parameters in accordance with the original dictates of British wool as applied today by the Australian Wool Exchange (AWEX) Council. Only a few dozen of the millions of bales auctioned every year can be classified and marked 1PP.

In the United States, three classifications of wool are named in the Wool Products Labeling Act of 1939. "Wool" is "the fiber from the fleece of the sheep or lamb or hair of the Angora or Cashmere goat (and may include the so-called specialty fibers from the hair of the camel, alpaca, llama, and vicuna) which has never been reclaimed from any woven or felted wool product". "Virgin wool" and "new wool" are also used to refer to such never used wool. There are two categories of recycled wool (also called reclaimed or shoddy wool). "Reprocessed wool" identifies "wool which has been woven or felted into a wool product and subsequently reduced to a fibrous state without having been used by the ultimate consumer". "Reused wool" refers to such wool that *has* been used by the ultimate consumer.

History

Wild sheep were more hairy than woolly. Although sheep were domesticated some 9,000 to 11,000 years ago, archaeological evidence from statuary found at sites in Iran suggests selection for woolly sheep may have begun around 6000 BC, with the earliest woven wool garments having only been dated to two to three thousand years later. Woolly-sheep were introduced into Europe from the Near East in the early part of the 4th millennium BC. The oldest known European wool textile, ca. 1500 BC, was preserved in a Danish bog. Prior to invention of shears—probably in the Iron Age—the wool was plucked out by hand or by bronze combs. In Roman times, wool, linen, and leather clothed the European population; cotton from India was a curiosity of which only naturalists had heard, and silks, imported along the Silk Road from China, were extravagant luxury goods. Pliny the Elder records in his *Natural History* that the reputation for producing the finest wool was enjoyed by Tarentum, where selective breeding had produced sheep with superior fleeces, but which required special care.

A 1905 illustration of a Tibetan man spinning wool

In medieval times, as trade connections expanded, the Champagne fairs revolved around the production of wool cloth in small centers such as Provins. The network developed by the annual fairs meant the woolens of Provins might find their way to Naples, Sicily, Cyprus, Majorca, Spain, and even Constantinople. The wool trade developed into serious business, a generator of capital. In the 13th century, the wool trade became the economic engine of the Low Countries and central Italy. By the end of the 14th century, Italy predominated, though Italian production turned to silk in the 16th century. Both industries, based on the export of English raw wool, were rivaled only by the 15th-century sheepwalks of Castile and were a significant source of income to the English crown, which in 1275 had imposed an export tax on wool called the "Great Custom". The importance of wool to the English economy can be seen in the fact that since the 14th century, the presiding officer of the House of Lords has sat on the "Woolsack", a chair stuffed with wool.

Economies of scale were instituted in the Cistercian houses, which had accumulated great tracts of land during the 12th and early 13th centuries, when land prices were low and labor still scarce. Raw wool was baled and shipped from North Sea ports to the textile cities of Flanders, notably Ypres and Ghent, where it was dyed and worked up as cloth. At the time of the Black Death, English textile industries accounted for about 10% of English wool production; the English textile trade grew during the 15th century, to the point where export of wool was discouraged. Over the centuries, various British laws controlled the wool trade or required the use of wool even in burials. The smuggling of wool out of the country, known as owling, was at one time punishable by the cutting off of a hand. After the Restoration, fine English woolens began to compete with silks in the international market, partly aided by the Navigation Acts; in 1699, the English crown forbade its American colonies to trade wool with anyone but England herself.

A great deal of the value of woolen textiles was in the dyeing and finishing of the woven product. In each of the centers of the textile trade, the manufacturing process came to be subdivided into a collection of trades, overseen by an entrepreneur in a system called by the English the "putting-out" system, or "cottage industry", and the *Verlagssystem* by the Germans. In this system of producing wool cloth, until recently perpetuated in the production of Harris tweeds, the entrepreneur provides the raw materials and an advance, the remainder being paid upon delivery of the product. Written contracts bound the artisans to specified terms. Fernand Braudel traces the appearance of the system in the 13th-century economic boom, quoting a document of 1275. The system effectively bypassed the guilds' restrictions.

Before the flowering of the Renaissance, the Medici and other great banking houses of Florence had built their wealth and banking system on their textile industry based on wool, overseen by the Arte della Lana, the wool guild: wool textile interests guided Florentine policies. Francesco Datini, the "merchant of Prato", established in 1383 an *Arte della Lana* for that small Tuscan city. The sheepwalks of Castile shaped the landscape and the fortunes of the *meseta* that lies in the heart of the Iberian peninsula; in the 16th century, a unified Spain allowed export of Merino lambs only with royal permission. The German wool market – based on sheep of Spanish origin – did not overtake British wool until comparatively late. The Industrial Revolution introduced mass production technology into wool and wool cloth manufacturing. Australia's colonial economy was based on sheep raising, and the Australian wool trade eventually overtook that of the Germans by 1845, furnishing wool for Bradford, which developed as the heart of industrialized woolens production.

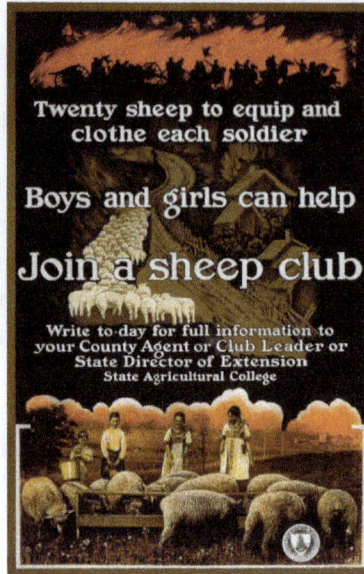

A World War I-era poster sponsored by the United States Department of Agriculture encouraging children to raise sheep to provide needed war supplies

Due to decreasing demand with increased use of synthetic fibers, wool production is much less than what it was in the past. The collapse in the price of wool began in late 1966 with a 40% drop; with occasional interruptions, the price has tended down. The result has been sharply reduced production and movement of resources into production of other commodities, in the case of sheep growers, to production of meat.

Superwash wool (or washable wool) technology first appeared in the early 1970s to produce wool that has been specially treated so it is machine washable and may be tumble-dried. This wool is produced using an acid bath that removes the "scales" from the fiber, or by coating the fiber with a polymer that prevents the scales from attaching to each other and causing shrinkage. This process results in a fiber that holds longevity and durability over synthetic materials, while retaining its shape.

In December 2004, a bale of the then world's finest wool, averaging 11.8 microns, sold for AU$3,000 per kilogram at auction in Melbourne, Victoria. This fleece wool tested with an average yield of 74.5%, 68 mm long, and had 40 newtons per kilotex strength. The result was A$279,000 for the bale. The finest bale of wool ever auctioned was sold for a seasonal record of AU$2690 per kilo during June 2008. This bale was produced by the Hillcreston Pinehill Partnership and measured 11.6 microns, 72.1% yield, and had a 43 newtons per kilotex strength measurement. The bale realized $247,480 and was exported to India.

In 2007, a new wool suit was developed and sold in Japan that can be washed in the shower, and which dries off ready to wear within hours with no ironing required. The suit was developed using Australian Merino wool, and it enables woven products made from wool, such as suits, trousers, and skirts, to be cleaned using a domestic shower at home.

In December 2006, the General Assembly of the United Nations proclaimed 2009 to be the International Year of Natural Fibres, so as to raise the profile of wool and other natural fibers.

Production

Global wool production is about 2 million tonnes per year, of which 60% goes into apparel. Australia is a leading producer of wool which is mostly from Merino sheep, but has recently been eclipsed by China in terms of total weight. New Zealand is now (2016) the third-largest producer of wool, and the largest producer of crossbred wool. Breeds such as Lincoln, Romney, Drysdale, and Elliotdale produce coarser fibers, and wool from these sheep is usually used for making carpets.

In the United States, Texas, New Mexico, and Colorado have large commercial sheep flocks and their mainstay is the Rambouillet (or French Merino). Also, a thriving home-flock contingent of small-scale farmers raise small hobby flocks of specialty sheep for the hand-spinning market. These small-scale farmers offer a wide selection of fleece. Global woolclip (total amount of wool shorn) 2004/2005

1. Australia: 25% of global woolclip (475 million kg greasy, 2004/2005)
2. China: 18%
3. United States: 17%
4. New Zealand: 11%
5. Argentina: 3%
6. Turkey: 2%
7. Iran: 2%
8. United Kingdom: 2%
9. India: 2%
10. Sudan: 2%
11. South Africa: 1%

Organic wool is becoming more and more popular. This wool is very limited in supply and much of it comes from New Zealand and Australia. It is becoming easier to find in clothing and other products, but these products often carry a higher price. Wool is environmentally preferable (as compared to petroleum-based nylon or polypropylene) as a material for carpets, as well, in particular when combined with a natural binding and the use of formaldehyde-free glues.

Animal rights groups have noted issues with the production of wool, such as mulesing.

Marketing

Australia

About 85% of wool sold in Australia is sold by open cry auction. "Sale by sample" is a method in which a mechanical claw takes a sample from each bale in a line or lot of wool. These grab samples are bulked, objectively measured, and a sample of not less than 4 kg is displayed in a box for the buyer to examine. The Australian Wool Exchange conducts sales primarily in Sydney, Melbourne, Newcastle, and Fremantle. About 80 brokers and agents work throughout Australia.

"Wool: Fibre of the gods, created – not man-made" CSIRO marketing poster describing the benefits of wool

Merino wool samples for sale by auction, Newcastle, New South Wales

Wool received by Australian brokers and dealers (tonnes/quarter) since 1973

Wool buyers' room at a wool auction, Newcastle, New South Wales

About 7% of Australian wool is sold by private treaty on farms or to local wool-handling facilities. This option gives wool growers benefit from reduced transport, warehousing, and selling costs. This method is preferred for small lots or mixed butts to make savings on reclassing and testing.

About 5% of Australian wool is sold over the internet on an electronic offer board. This option gives wool growers the ability to set firm price targets, reoffer passed-in wool, and offer lots to the market quickly and efficiently. This method works well for tested lots, as buyers use these results to make a purchase. About 97% of wool is sold without sample inspection; however, as of December 2009, 59% of wool listed had been passed in from auction. Growers through certain brokers can allocate their wool to a sale and at what price their wool will be reserved.

Sale by tender can achieve considerable cost savings on wool clips large enough to make it worthwhile for potential buyers to submit tenders. Some marketing firms sell wool on a consignment basis, obtaining a fixed percentage as commission.

Forward selling: Some buyers offer a secure price for forward delivery of wool based on estimated measurements or the results of previous clips. Prices are quoted at current market rates and are locked in for the season. Premiums and discounts are added to cover variations in micron, yield, tensile strength, etc., which are confirmed by actual test results when available.

Another method of selling wool includes sales direct to wool mills.

Other Countries

The British Wool Marketing Board operates a central marketing system for UK fleece wool with the aim of achieving the best possible net returns for farmers.

Less than half of New Zealand's wool is sold at auction, while around 45% of farmers sell wool directly to private buyers and end-users.

United States sheep producers market wool with private or cooperative wool warehouses, but wool pools are common in many states. In some cases, wool is pooled in a local market area, but sold through a wool warehouse. Wool offered with objective measurement test results is preferred. Imported apparel wool and carpet wool goes directly to central markets, where it is handled by the large merchants and manufacturers.

Yarn

Virgin wool is spun for the first time.

Worsted wool yarn, the first step in the manufacture of most wool clothing.

Shoddy or recycled wool is made by cutting or tearing apart existing wool fabric and respinning the resulting fibers. As this process makes the wool fibers shorter, the remanufactured fabric is inferior to the original. The recycled wool may be mixed with raw wool, wool noil, or another fiber such as cotton to increase the average fiber length. Such yarns are typically used as weft yarns with a cotton warp. This process was invented in the Heavy Woollen District of West Yorkshire and created a microeconomy in this area for many years.

Rag is a sturdy wool fiber made into yarn and used in many rugged applications such as gloves.

Worsted is a strong, long-staple, combed wool yarn with a hard surface.

Woolen is a soft, short-staple, carded wool yarn typically used for knitting. In traditional weaving, woolen weft yarn (for softness and warmth) is frequently combined with a worsted warp yarn for strength on the loom.

Uses

In addition to clothing, wool has been used for blankets, horse rugs, saddle cloths, carpeting, felt, wool insulation and upholstery. Wool felt covers piano hammers, and it is used to absorb odors and noise in heavy machinery and stereo speakers. Ancient Greeks lined their helmets with felt, and Roman legionnaires used breastplates made of wool felt.

Woolen garments in the wool samples area of a wool store, Newcastle, New South Wales.

Wool has also been traditionally used to cover cloth diapers. Wool fiber exteriors are hydrophobic (repel water) and the interior of the wool fiber is hygroscopic (attracts water); this makes a wool garment suitable cover for a wet diaper by inhibiting wicking, so outer garments remain dry. Wool felted and treated with lanolin is water resistant, air permeable, and slightly antibacterial, so it resists the buildup of odor. Some modern cloth diapers use felted wool fabric for covers, and there are several modern commercial knitting patterns for wool diaper covers.

Initial studies of woolen underwear have found it prevented heat and sweat rashes because it more readily absorbs the moisture than other fibers.

Merino wool has been used in baby sleep products such as swaddle baby wrap blankets and infant sleeping bags.

As an animal protein, it can be used as a soil fertilizer, being a slow-release source of nitrogen.

Researchers at the Royal Melbourne Institute of Technology school of fashion and textiles have discovered a blend of wool and kevlar, the synthetic fiber widely used in body armor, was lighter, cheaper and worked better in damp conditions than kevlar alone. Kevlar, when used alone, loses about 20% of its effectiveness when wet, so required an expensive waterproofing process. Wool increased friction in a vest with 28–30 layers of fabric, to provide the same level of bullet resistance as 36 layers of Kevlar alone.

Carbon Footprint

The carbon footprint of woolen products is difficult to determine and varies considerably according to factors ranging from the conditions in which sheep are bred to the processes used to manufacture the products.

The average greenhouse gas footprint of wool in manufacturing carpets estimated at 5.48 kg CO_2 equivalent per kg, when produced in Europe.

The carbon footprint of a 264.85g woolly jumper made from New Zealand merino wool averages about 1.667 kg CO_2 equivalent at the point of purchase by a consumer, implying a carbon footprint of around 6 kg per kg of finished woolen product.

Events

A buyer of Merino wool, Ermenegildo Zegna, has offered awards for Australian wool producers. In 1963, the first Ermenegildo Zegna Perpetual Trophy was presented in Tasmania for growers of "Superfine skirted Merino fleece". In 1980, a national award, the Ermenegildo Zegna Trophy for Extrafine Wool Production, was launched. In 2004, this award became known as the Ermenegildo Zegna Unprotected Wool Trophy. In 1998, an Ermenegildo Zegna Protected Wool Trophy was launched for fleece from sheep coated for around nine months of the year.

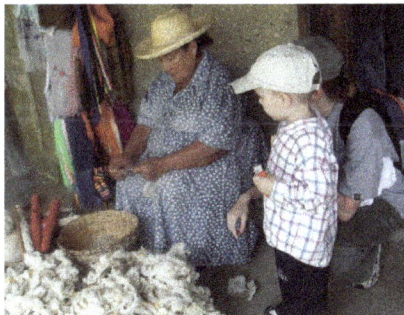

Andean woman sorting wool as part of the theme park Los Aleros in Mérida, Venezuela

In 2002, the Ermenegildo Zegna Vellus Aureum Trophy was launched for wool that is 13.9 microns or finer. Wool from Australia, New Zealand, Argentina, and South Africa may enter, and a winner is named from each country. In April 2008, New Zealand won the Ermenegildo Zegna Vellus Aureum Trophy for the first time with a fleece that measured 10.8 microns. This contest awards the winning fleece weight with the same weight in gold as a prize, hence the name.

In 2010, an ultrafine, 10-micron fleece, from Windradeen, near Pyramul, New South Wales, won the Ermenegildo Zegna Vellus Aureum International Trophy.

Since 2000, Loro Piana has awarded a cup for the world's finest bale of wool that produces just enough fabric for 50 tailor-made suits. The prize is awarded to an Australian or New Zealand wool grower who produces the year's finest bale.

The New England Merino Field days which display local studs, wool, and sheep are held during January, in even numbered years around the Walcha, New South Wales district. The Annual Wool Fashion Awards, which showcase the use of Merino wool by fashion designers, are hosted by the city of Armidale, New South Wales, in March each year. This event encourages young and established fashion designers to display their talents. During each May, Armidale hosts the annual New England Wool Expo to display wool fashions, handicrafts, demonstrations, shearing competitions, yard dog trials, and more.

In July, the annual Australian Sheep and Wool Show is held in Bendigo, Victoria. This is the largest sheep and wool show in the world, with goats and alpacas, as well as woolcraft competitions and displays, fleece competitions, sheepdog trials, shearing, and wool handling. The largest competition in the world for objectively measured fleeces is the Australian Fleece Competition, which is held annually at Bendigo. In 2008, 475 entries came from all states of Australia, with first and second prizes going to the Northern Tablelands, New South Wales fleeces.

Honey

Honey /ˈhʌni/ is a sweet food made by bees foraging nectar from flowers. The variety produced by honey bees (the genus *Apis*) is the one most commonly referred to, as it is the type of honey collected by most beekeepers and consumed by people. Honey is also produced by bumblebees, stingless bees, and other hymenopteran insects such as honey wasps, though the quantity is generally lower and they have slightly different properties compared to honey from the genus *Apis*. Honey bees convert nectar into honey by a process of regurgitation and evaporation: they store it as a primary food source in wax honeycombs inside the beehive.

Honey gets its sweetness from the monosaccharides fructose and glucose, and has about the same relative sweetness as granulated sugar. It has attractive chemical properties for baking and a distinctive flavor that leads some people to prefer it to sugar and other sweeteners. Most microorganisms do not grow in honey so sealed honey does not spoil, even after thousands of years. However, honey sometimes contains dormant endospores of the bacterium *Clostridium botulinum*, which can be dangerous to babies, as it may result in botulism.

A jar of honey with a honey dipper and an American biscuit

People who have a weakened immune system should not eat honey because of the risk of bacterial or fungal infection. There is some evidence that honey may be effective in treating diseases and other medical conditions such as wounds and burns. However, the evidence is overall not conclusive. Providing 64 calories in a typical serving of one tablespoon (15 mL), honey contains no significant essential nutrient content. Honey is generally safe but there are various, potential adverse-effects or interactions it may have in combination with excessive consumption, existing disease conditions, or drugs.

Honey in honeycomb

Honey use and production has a long and varied history. Honey collection is an ancient activity. Humans apparently began foraging for honey at least 8,000 years ago, as evidenced by a cave painting in Valencia, Spain.

Formation

Honey is produced by bees from nectar collection which serves the dual purpose to support metabolism of muscle activity during foraging and for long-term food storage as honey. During foraging, bees access part of the nectar collected to support metabolic activity of flight muscles, with the majority of collected nectar destined for regurgitation, digestion and storage as honey. In cold weather or when other food sources are scarce, adult and larval bees use stored honey as food.

A honey bee on calyx of goldenrod

By contriving for bee swarms to nest in artificial hives, people have been able to semi-domesticate the insects and harvest excess honey. In the hive or in a wild nest, the three types of bees are:

- a single female queen bee

- a seasonally variable number of male drone bees to fertilize new queens

- 20,000 to 40,000 female worker bees

Leaving the hive, foraging bees collect sugar-rich flower nectar and return to the hive where they use their "honey stomachs" to ingest and regurgitate the nectar repeatedly until it is partially digested. Bee digestive enzymes - invertase, amylase and diastase - and gastric acid hydrolyze sucrose to a mixture of glucose and fructose. The bees work together as a group with the regurgitation and digestion for as long as 20 minutes until the product reaches storage quality. It is then placed in honeycomb cells left unsealed while still high in water content (about 20%) and natural yeasts, which, unchecked, would cause the sugars in the newly formed honey to ferment. The process continues as hive bees flutter their wings constantly to circulate air and evaporate water from the honey to a content of about 18%, raising the sugar concentration and preventing fermentation. The bees then cap the cells with wax to seal them. As removed from the hive by a beekeeper, honey has a long shelf life and will not ferment if properly sealed.

Another source of honey is from a number of wasp species, such as the wasps *Brachygastra lecheguana* and *Brachygastra mellifica*, which are found in South and Central America. These species are known to feed on nectar and produce honey.

Some wasps, such as the *Polistes versicolor*, even consume honey themselves, switching from feeding on pollen in the middle of their lifecycles to feeding on honey, which can better provide for their energy needs.

Extraction from a honeycomb

Filtering from a honeycomb

Pouring raw honey

Collection

Honey is collected from wild bee colonies, or from domesticated beehives. Wild bee nests are sometimes located by following a honeyguide bird. The bees may first be pacified by using smoke from a bee smoker. The smoke triggers a feeding instinct (an attempt to save the resources of the hive from a possible fire), making them less aggressive and the smoke obscures the pheromones the bees use to communicate.

The honeycomb is removed from the hive and the honey may be extracted from that, either by crushing or by using a honey extractor. The honey is then usually filtered to remove beeswax and other debris.

Before the invention of removable frames, bee colonies were often sacrificed in order to conduct the harvest. The harvester would take all the available honey and replace the entire colony the next spring. Since the invention of removable frames, the principles of husbandry lead most beekeepers to ensure that their bees will have enough stores to survive the winter, either by leaving some honey in the beehive or by providing the colony with a honey substitute such as sugar water or crystalline sugar (often in the form of a "candyboard"). The amount of food necessary to survive the winter depends on the variety of bees and on the length and severity of local winters.

A wide range of species other than humans are attracted to wild or domestic sources of honey.

Production

Top five honey producing countries (millions of tonnes)			
Rank	**Country**	**2013**	
1	China	0.47	
2	Turkey	0.09	
3	Argentina	0.08	
4	Ukraine	0.07	
5	Russia	0.07	
--	**World**	**1.7**	
Source: UN Food & Agriculture Organization, FAOSTAT			

In 2013, 1.7 million tonnes of honey were produced worldwide, with China accounting for 28% of the world total (table). The next four largest producers – Turkey, Argentina, Ukraine and Russia – accounted collectively for less than 20% of the world total (table).

Modern Uses

Food

Over its history as a food, the main uses of honey are in cooking, baking, desserts, such as *mel i mató*, as a spread on bread, and as an addition to various beverages, such as tea, and as a sweetener in some commercial beverages. Honey barbecue and honey mustard are other common flavors used in sauces.

Fermentation

Honey is the main ingredient in the alcoholic beverage mead, which is also known as "honey wine" or "honey beer". Historically, the ferment for mead was honey's naturally occurring yeast. Honey is also used as an adjunct in some beers.

Honey wine, or mead, is typically (modern era) made with a honey and water mixture with yeast added for fermentation. Primary fermentation usually takes 28–56 days, after which the must needs to be racked into a secondary fermentation vessel and left to sit about 35–40 more days. If done properly, fermentation will be finished by this point (though if a sparkling mead is desired, fermentation can be restarted after bottling by the addition of a small amount of sugar), but most meads require aging for 6–9 months or more in order to be palatable.

Physical and Chemical Properties

The physical properties of honey vary, depending on water content, the type of flora used to produce it (pasturage), temperature, and the proportion of the specific sugars it contains. Fresh honey is a supersaturated liquid, containing more sugar than the water can typically dissolve at ambient temperatures. At room temperature, honey is a supercooled liquid, in which the glucose will precipitate into solid granules. This forms a semisolid solution of precipitated glucose crystals in a solution of fructose and other ingredients.

Crystallized honey. The inset shows a close-up of the honey, showing the individual glucose grains in the fructose mixture.

Phase Transitions

The melting point of crystallized honey is between 40 and 50 °C (104 and 122 °F), depending on its composition. Below this temperature, honey can be either in a metastable state, meaning that it will not crystallize until a seed crystal is added, or, more often, it is in a "labile" state, being saturated with enough sugars to crystallize spontaneously. The rate of crystallization is affected by many factors, but the primary factor is the ratio of the main sugars: fructose to glucose. Honeys that are supersaturated with a very high percentage of glucose, such as brassica honey, will crystallize almost immediately after harvesting, while honeys with a low percentage of glucose, such as chestnut or tupelo honey, do not crystallize. Some types of honey may produce very large but few crystals, while others will produce many small crystals.

Crystallization is also affected by water content, because a high percentage of water will inhibit crystallization, as will a high dextrin content. Temperature also affects the rate of crystallization, with the fastest growth occurring between 13 and 17 °C (55 and 63 °F).

Crystal nuclei (seeds) tend to form more readily if the honey is disturbed, by stirring, shaking or agitating, rather than if left at rest. However, the nucleation of microscopic seed-crystals is greatest between 5 and 8 °C (41 and 46 °F). Therefore, larger but fewer crystals tend to form at higher temperatures, while smaller but more-numerous crystals usually form at lower temperatures. Below 5 °C, the honey will not crystallize and, thus, the original texture and flavor can be preserved indefinitely.

Since honey normally exists below its melting point, it is a supercooled liquid. At very low temperatures, honey will not freeze solid. Instead, as the temperatures become lower, the viscosity of honey increases. Like most viscous liquids, the honey will become thick and sluggish with decreasing temperature. At −20 °C (−4 °F), honey may appear or even feel solid, but it will continue to flow at very low rates. Honey has a glass transition between −42 and −51 °C (−44 and −60 °F). Below this temperature, honey enters a glassy state and will become an amorphous solid (noncrystalline).

Viscosity

The viscosity of honey is affected greatly by both temperature and water content. The higher the water percentage, the easier honey flows. Above its melting point, however, water has little effect on viscosity. Aside from water content, the composition of honey also has little effect on viscosity, with the exception of a few types. At 25 °C (77 °F), honey with 14% water content generally has a viscosity around 400 poise, while a honey containing 20% water has a viscosity around 20 poise. Viscosity increase due to temperature occurs very slowly at first. A honey containing 16% water, at 70 °C (158 °F), will have a viscosity around 2 poise, while at 30 °C (86 °F), the viscosity is around 70 poise. As cooling progresses, honey becomes more viscous at an increasingly rapid rate, reaching 600 poise around 14 °C (57 °F). However, while honey is very viscous, it has rather low surface tension.

A few types of honey have unusual viscous properties. Honeys from heather or manuka display thixotropic properties. These types of honey enter a gel-like state when motionless, but then liquify when stirred.

Electrical and Optical Properties

Because honey contains electrolytes, in the form of acids and minerals, it exhibits varying degrees of electrical conductivity. Measurements of the electrical conductivity are used to determine the quality of honey in terms of ash content.

The effect honey has on light is useful for determining the type and quality. Variations in the water content alter the refractive index of honey. Water content can easily be measured with a refractometer. Typically, the refractive index for honey will range from 1.504 at 13% water content to 1.474 at 25%. Honey also has an effect on polarized light, in that it will rotate the polarization plane. The fructose will give a negative rotation, while the glucose will give a positive one. The overall rotation can be used to measure the ratio of the mixture. Honey may vary in color between pale yellow and dark brown, but other bright colors may occasionally be found, depending on the source of the sugar harvested by the bees.

Hygroscopy and Fermentation

Honey has the ability to absorb moisture directly from the air, a phenomenon called hygroscopy. The amount of water the honey will absorb is dependent on the relative humidity of the air. Because honey contains yeast, this hygroscopic nature requires that honey be stored in sealed containers to prevent fermentation, which usually begins if the honey's water content rises much above 25%. Honey will tend to absorb more water in this manner than the individual sugars would allow on their own, which may be due to other ingredients it contains.

Fermentation of honey will usually occur after crystallization because, without the glucose, the liquid portion of the honey primarily consists of a concentrated mixture of the fructose, acids, and water, providing the yeast with enough of an increase in the water percentage for growth. Honey that is to be stored at room temperature for long periods of time is often pasteurized, to kill any yeast, by heating it above 70 °C (158 °F).

Thermal Characteristics

Like all sugar compounds, honey will caramelize if heated sufficiently, becoming darker in color, and eventually burn. However, honey contains fructose, which caramelizes at lower temperatures than the glucose. The temperature at which caramelization begins varies, depending on the composition, but is typically between 70 and 110 °C (158 and 230 °F). Honey also contains acids, which act as catalysts, decreasing the caramelization temperature even more. Of these acids, the amino acids, which occur in very small amounts, play an important role in the darkening of honey. The amino acids form darkened compounds called melanoidins, during a Maillard reaction. The Maillard reaction will occur slowly at room temperature, taking from a few to several months to show visible darkening, but will speed-up dramatically with increasing temperatures. However, the reaction can also be slowed by storing the honey at colder temperatures.

Unlike many other liquids, honey has very poor thermal conductivity, taking a long time to reach thermal equilibrium. Melting crystallized honey can easily result in localized caramelization if the heat source is too hot, or if it is not evenly distributed. However, honey will take substantially longer to liquify when just above the melting point than it will at elevated temperatures. Melting 20 kilograms of crystallized honey, at 40 °C (104 °F), can take up to 24 hours, while 50 kilograms may take twice as long. These times can be cut nearly in half by heating at 50 °C (122 °F). However, many of the minor substances in honey can be affected greatly by heating, changing the flavor, aroma, or other properties, so heating is usually done at the lowest temperature possible for the shortest amount of time.

Classification

Honey is classified by its floral source, and there are also divisions according to the packaging and processing used. There are also regional honeys. In the USA honey is also graded on its color and optical density by USDA standards, graded on the Pfund scale, which ranges from 0 for "water white" honey to more than 114 for "dark amber" honey.

Honey

Floral Source

Generally, honey is classified by the floral source of the nectar from which it was made. Honeys can be from specific types of flower nectars or can be blended after collection. The pollen in honey is traceable to floral source and therefore region of origin. The rheological and melissopalynological properties of honey can be used to identify the major plant nectar source used in its production.

Blended

Most commercially available honey is blended, meaning it is a mixture of two or more honeys differing in floral source, color, flavor, density or geographic origin.

Polyfloral

Polyfloral honey, also known as wildflower honey, is derived from the nectar of many types of flowers.

The taste may vary from year to year, and the aroma and the flavor can be more or less intense, depending on which bloomings are prevalent.

Monofloral

Monofloral honey is made primarily from the nectar of one type of flower. Different monofloral honeys have a distinctive flavor and color because of differences between their principal nectar sources. To produce monofloral honey, beekeepers keep beehives in an area where the bees have access to only one type of flower. In practice, because of the difficulties in containing bees, a small proportion of any honey will be from additional nectar from other flower types. Typical examples of North American monofloral honeys are clover, orange blossom, blueberry, sage, tupelo, buckwheat, fireweed, mesquite and sourwood. Some typical European examples include thyme, thistle, heather, acacia, dandelion, sunflower, lavender, honeysuckle, and varieties from lime and chestnut trees. In North Africa (e.g. Egypt) examples include clover, cotton, and citrus (mainly orange blossoms). The unique flora of Australia yields a number of distinctive honeys, with some of the most popular being yellow box, blue gum, ironbark, bush mallee, Tasmanian leatherwood, and macadamia.

Honeydew Honey

Instead of taking nectar, bees can take honeydew, the sweet secretions of aphids or other plant sap-sucking insects. Honeydew honey is very dark brown in color, with a rich fragrance of stewed fruit or fig jam, and is not as sweet as nectar honeys. Germany's Black Forest is a well known source of honeydew-based honeys, as well as some regions in Bulgaria, Tara (mountain) in Serbia and Northern California in the United States. In Greece, pine honey (a type of honeydew honey) constitutes 60–65% of the annual honey production. Honeydew honey is popular in some areas, but in other areas beekeepers have difficulty selling the stronger flavored product.

The production of honeydew honey has some complications and dangers. The honey has a much larger proportion of indigestibles than light floral honeys, thus causing dysentery to the bees, resulting in the death of colonies in areas with cold winters. Good beekeeping management requires the removal of honeydew prior to winter in colder areas. Bees collecting this resource also have to be fed protein supplements, as honeydew lacks the protein-rich pollen accompaniment gathered from flowers.

Classification by Packaging and Processing

Generally, honey is bottled in its familiar liquid form. However, honey is sold in other forms, and can be subjected to a variety of processing methods.

Honeycomb

A variety of honey flavors and container sizes and styles from the 2008 Texas State Fair

- Crystallized honey is honey in which some of the glucose content has spontaneously crystallized from solution as the monohydrate. Also called "granulated honey" or "candied honey." Honey that has crystallized (or commercially purchased crystallized) can be returned to a liquid state by warming.

- Pasteurized honey is honey that has been heated in a pasteurization process which requires temperatures of 161 °F (72 °C) or higher. Pasteurization destroys yeast cells. It also liquefies any microcrystals in the honey, which delays the onset of visible crystallization. However, excessive heat exposure also results in product deterioration, as it increases the level of hydroxymethylfurfural (HMF) and reduces enzyme (e.g. diastase) activity. Heat also affects appearance (darkens the natural honey color), taste, and fragrance.

- Raw honey is honey as it exists in the beehive or as obtained by extraction, settling or straining, without adding heat (although some honey that has been "minimally processed" is often labeled as raw honey). Raw honey contains some pollen and may contain small particles of wax.

- Strained honey has been passed through a mesh material to remove particulate material (pieces of wax, propolis, other defects) without removing pollen, minerals or enzymes.

- Filtered honey is honey of any type that has been filtered to the extent that all or most of the fine particles, pollen grains, air bubbles, or other materials normally found in suspension, have been removed. The process typically heats honey to 150–170 °F (66–77 °C) to more easily pass through the filter. Filtered honey is very clear and will not crystallize as quickly, making it preferred by the supermarket trade.

- Ultrasonicated honey has been processed by ultrasonication, a non-thermal processing alternative for honey. When honey is exposed to ultrasonication, most of the yeast cells are destroyed. Those cells that survive sonication generally lose their ability to grow, which reduces the rate of honey fermentation substantially. Ultrasonication also eliminates existing crystals and inhibits further crystallization in honey. Ultrasonically aided liquefaction can work at substantially lower temperatures of approximately 95 °F (35 °C) and can reduce liquefaction time to less than 30 seconds.

- Creamed honey, also called whipped honey, spun honey, churned honey, honey fondant, and (in the UK) set honey, has been processed to control crystallization. Creamed honey contains a large number of small crystals, which prevent the formation of larger crystals that can occur in unprocessed honey. The processing also produces a honey with a smooth, spreadable consistency.

- Dried honey has the moisture extracted from liquid honey to create completely solid, nonsticky granules. This process may or may not include the use of drying and anticaking agents. Dried honey is used in baked goods, and to garnish desserts.

- Comb honey is honey still in the honeybees' wax comb. It is traditionally collected by using standard wooden frames in honey supers. The frames are collected and the comb is cut out in chunks before packaging. As an alternative to this labor-intensive method, plastic rings or cartridges can be used that do not require manual cutting of the comb, and speed packaging. Comb honey harvested in the traditional manner is also referred to as "cut-comb honey".

- Chunk honey is packed in widemouth containers consisting of one or more pieces of comb honey immersed in extracted liquid honey.

- Honey decoctions are made from honey or honey by-products which have been dissolved in water, then reduced (usually by means of boiling). Other ingredients may then be added. (For example, abbamele has added citrus.) The resulting product may be similar to molasses.

- Baker's honey is honey which is outside the normal specification for honey, due to a "foreign" taste or odor, or because it has begun to ferment or has been overheated. It is generally used as an ingredient in food processing. There are additional requirements for labeling baker's honey, including that it may not be sold labelled simply as "honey".

Grading

In the US, honey grading is performed voluntarily (USDA does offer inspection and grading "as on-line (in-plant) or lot inspection...upon application, on a fee-for-service basis.") based upon USDA standards. Honey is graded based upon a number of factors, including water content, flavor and aroma, absence of defects and clarity. Honey is also classified by color though it is not a factor in the grading scale. The honey grade scale is:

Grade	Soluble solids	Flavor and aroma	Absence of defects	Clarity
A	≥ 81.4%	Good—"has a good, normal flavor and aroma for the predominant floral source or, when blended, a good flavor for the blend of floral sources and the honey is free from caramelized flavor or objectionable flavor caused by fermentation, smoke, chemicals, or other causes with the exception of the predominant floral source"	Practically free—"contains practically no defects that affect the appearance or edibility of the product"	Clear—"may contain air bubbles which do not materially affect the appearance of the product and may contain a trace of pollen grains or other finely divided particles of suspended material which do not affect the appearance of the product"
B	≥ 81.4%	Reasonably good—"has a reasonably good, normal flavor and aroma for the predominant floral source or, when blended, a reasonably good flavor for the blend of floral sources and the honey is practically free from caramelized flavor and is free from objectionable flavor caused by fermentation, smoke, chemicals, or other causes with the exception of the predominant floral source"	Reasonably free—"may contain defects which do not materially affect the appearance or edibility of the product"	Reasonably clear—"may contain air bubbles, pollen grains, or other finely divided particles of suspended material which do not materially affect the appearance of the product"
C	≥ 80.0%	Fairly good—"has a fairly good, normal flavor and aroma for the predominant floral source or, when blended, a fairly good flavor for the blend of floral sources and the honey is reasonably free from caramelized flavor and is free from objectionable flavor caused by fermentation, smoke, chemicals, or other causes with the exception of the predominant floral source"	Fairly free—"may contain defects which do not seriously affect the appearance or edibility of the product"	Fairly clear—"may contain air bubbles, pollen grains, or other finely divided particles of suspended material which do not seriously affect the appearance of the product"
Substandard	Fails Grade C	Fails Grade C	Fails Grade C	Fails Grade C

Other countries may have differing standards on the grading of honey. India, for example, certifies honey grades based on additional factors, such as the Fiehe's test, and other empirical measurements.

Indicators of Quality

High-quality honey can be distinguished by fragrance, taste, and consistency. Ripe, freshly collected, high-quality honey at 20 °C (68 °F) should flow from a knife in a straight stream, without breaking into separate drops. After falling down, the honey should form a bead. The honey, when poured, should form small, temporary layers that disappear fairly quickly, indicating high viscosity. If not, it indicates excessive water content (over 20%) of the product. Honey with excessive water content is not suitable for long-term preservation.

In jars, fresh honey should appear as a pure, consistent fluid, and should not set in layers. Within a few weeks to a few months of extraction, many varieties of honey crystallize into a cream-colored solid. Some varieties of honey, including tupelo, acacia, and sage, crystallize less regularly. Honey may be heated during bottling at temperatures of 40–49 °C (104–120 °F) to delay or inhibit crystallization. Overheating is indicated by change in enzyme levels, for instance, diastase activity, which can be determined with the Schade or the Phadebas methods. A fluffy film on the surface of the honey (like a white foam), or marble-colored or white-spotted crystallization on a container's sides, is formed by air bubbles trapped during the bottling process.

A 2008 Italian study determined nuclear magnetic resonance spectroscopy can be used to distinguish between different honey types, and can be used to pinpoint the area where it was produced. Researchers were able to identify differences in acacia and polyfloral honeys by the differing proportions of fructose and sucrose, as well as differing levels of aromatic amino acids phenylalanine and tyrosine. This ability allows greater ease of selecting compatible stocks.

Acid Content and Flavor Effects

The average pH of honey is 3.9, but can range from 3.4 to 6.1. Honey contains many kinds of acids, both organic and amino. However, the different types and their amounts vary considerably, depending on the type of honey. These acids may be aromatic or aliphatic (non-aromatic). The aliphatic acids contribute greatly to the flavor of honey by interacting with the flavors of other ingredients.

Organic acids comprise most of the acids in honey, accounting for 0.17–1.17% of the mixture, with gluconic acid formed by the actions of an enzyme called glucose oxidase as the most prevalent. Other organic acids are minor, consisting of formic, acetic, butyric, citric, lactic, malic, pyroglutamic, propionic, valeric, capronic, palmitic, and succinic, among many others.

Preservation

Because of its unique composition and chemical properties, honey is suitable for long-term storage, and is easily assimilated even after long preservation. Honey, and objects immersed in honey, have been preserved for centuries. The key to preservation is limiting access to humidity. In its cured state, honey has a sufficiently high sugar content to inhibit fermentation. If exposed to moist air, its hydrophilic properties will pull moisture into the honey, eventually diluting it to the point that fermentation can begin.

Sealed frame of honey

Regardless of preservation, honey may crystallize over time. The crystals can be dissolved by heating the honey.

Nutritional and Sugar Profile

In a 100 gram serving, honey provides 304 calories with no essential nutrients in significant content (table). Composed of 17% water and 82% carbohydrates, honey has low content of fat, dietary fiber and protein (table).

A mixture of sugars and other carbohydrates, honey is mainly fructose (about 38-55%) and glucose (about 31%), with remaining sugars including maltose, sucrose, and other complex carbohydrates. Its glycemic index ranges from 31 to 78, depending on the variety. The specific composition, color, aroma and flavor of any batch of honey depend on the flowers foraged by bees that produced the honey.

One 1980 study found that mixed floral honey from several United States regions typically contains:

- Fructose: 38.2%

- Glucose: 31.3%

- Maltose: 7.1%

- Sucrose: 1.3%

- Water: 17.2%

- Higher sugars: 1.5%

- Ash: 0.2%

- Other/undetermined: 3.2%

A 2013 NMR spectroscopy study of 20 different honeys from Germany found that their sugar contents comprised:

- Fructose: 28% to 41%

- Glucose: 22% to 35%

The average ratio was 56% fructose to 44% glucose, but the ratios in the individual honeys ranged from a high of 64% fructose and 36% glucose (one type of flower honey; Table 3 in reference) to a low of 50% fructose and 50% glucose (a different floral source). This NMR method was not able to quantify maltose, galactose, and the other minor sugars as compared to fructose and glucose.

Honey has a density of about 1.36 kilograms per litre (36% denser than water).

Adulteration

Adulteration of honey is the addition of other sugars, syrups or compounds into honey to change its flavor, viscosity, make it cheaper to produce, or to increase the fructose content in order to stave off crystallization. According to the Codex Alimentarius of the United Nations, any product labeled as honey or pure honey must be a wholly natural product, although different nations have their own laws concerning labeling. Adulteration of honey is sometimes used as a method of deception when buyers are led to believe that the honey is pure. The practice was common dating back to ancient times, when crystallized honey was often mixed with flour or other fillers, hiding the adulteration from buyers until the honey was liquefied. In modern times the most common adulteration-ingredient became clear, almost-flavorless corn syrup, which, when mixed with honey, is often very difficult to distinguish from unadulterated honey.

Isotope ratio mass spectrometry can be used to detect addition of corn syrup and cane sugar by the carbon isotopic signature. Addition of sugars originating from corn or sugar cane (C4 plants, unlike the plants used by bees, and also sugar beet, which are predominantly C3 plants) skews the isotopic ratio of sugars present in honey, but does not influence the isotopic ratio of proteins. In an unadulterated honey, the carbon isotopic ratios of sugars and proteins should match. Levels as low as 7% of addition can be detected.

In one country, the USA, according to The National Honey Board (a USDA-overseen organization), "honey stipulates a pure product that does not allow for the addition of any other substance...this includes, but is not limited to, water or other sweeteners".

Medical Uses

Wounds and Burns

Honey contains trace amount of compounds implicated in preliminary studies to have wound healing properties, such as hydrogen peroxide and methylglyoxal.

There is some evidence that honey may help healing in skin wounds after surgery and mild (partial thickness) burns when used in a dressing, but in general the evidence for the use of honey in wound treatment is of such low quality that firm conclusions cannot be drawn.

Evidence does not support the use of honey-based products in the treatment of venous stasis ulcers or ingrowing toenail.

There is ongoing research into medical uses for honey, particularly in the face of anti-microbial resistance to modern antibiotics.

Cough

For chronic cough and acute cough, a Cochrane review found no strong evidence for or against the use of honey. For treating children, the study concluded that honey possibly helps more than no treatment.

The UK Medicines and Healthcare Products Regulatory Agency recommends avoiding giving over the counter cough and common cold medication to children under 6, and suggests "a homemade remedy containing honey and lemon is likely to be just as useful and safer to take", but warns that honey should not be given to babies because of the risk of infant botulism. The World Health Organization recommends honey as a treatment for coughs and sore throats, including for children, stating that there is no reason to believe it is less effective than a commercial remedy. Honey is recommended by one Canadian physician for children over the age of 1 for the treatment of coughs as it is deemed as effective as dextromethorphan and more effective than diphenhydramine.

Other

People who have a weakened immune system should not eat honey because of the risk of bacterial or fungal infection.

No evidence shows the benefit of using honey to treat cancer, although honey may be useful for controlling side effects of radiation therapy or chemotherapy applied in cancer treatment.

Consumption is sometimes advocated as a treatment for seasonal allergies due to pollen, but there is inconclusive scientific evidence to support the claim. Honey is generally considered ineffective for the treatment of allergic conjunctivitis.

Preliminary studies found honey to contain an antimicrobial peptide called bee defensin-1. Some *in vitro* studies show that honey can kill Methicillin-resistant *Staphylococcus aureus* (MRSA), β-*haemolytic streptococci* and vancomycin-resistant *Enterococci*.

Health Hazards

Adverse Effects

Although honey is generally safe when taken in typical food amounts, there are various, potential adverse effects or interactions it may have in combination with excessive consumption, existing disease conditions or drugs. Included among these are mild reactions to high intake, such as anxiety, insomnia or hyperactivity in about 10% of children, according one study. No symptoms of anxiety, insomnia or hyperactivity were detected with honey consumption compared to placebo, according to another study. Honey consumption may interact adversely with existing allergies, high blood sugar levels (as in diabetes), or anticoagulants used to control bleeding, among other clinical conditions.

Botulism

Infants can develop botulism after consuming honey contaminated with *Clostridium botulinum* endospores.

Infantile botulism shows geographical variation. In the UK, only six cases have been reported between 1976 and 2006, yet the U.S. has much higher rates: 1.9 per 100,000 live births, 47.2% of which are in California. While the risk honey poses to infant health is small, it is recommended not to take the risk until after one year of age, and then giving honey is considered safe.

Toxic Honey

Mad honey intoxication is a result of eating honey containing grayanotoxins. Honey produced from flowers of rhododendrons, mountain laurels, sheep laurel, and azaleas may cause honey intoxication. Symptoms include dizziness, weakness, excessive perspiration, nausea, and vomiting. Less commonly, low blood pressure, shock, heart rhythm irregularities, and convulsions may occur, with rare cases resulting in death. Honey intoxication is more likely when using "natural" unprocessed honey and honey from farmers who may have a small number of hives. Commercial processing, with pooling of honey from numerous sources, is thought to dilute any toxins.

Toxic honey may also result when bees are proximate to tutu bushes (*Coriaria arborea*) and the vine hopper insect (*Scolypopa australis*). Both are found throughout New Zealand. Bees gather honeydew produced by the vine hopper insects feeding on the tutu plant. This introduces the poison tutin into honey. Only a few areas in New Zealand (the Coromandel Peninsula, Eastern Bay of Plenty and the Marlborough Sounds) frequently produce toxic honey. Symptoms of tutin poisoning include vomiting, delirium, giddiness, increased excitability, stupor, coma, and violent convulsions. To reduce the risk of tutin poisoning, humans should not eat honey taken from feral hives in the risk areas of New Zealand. Since December 2001, New Zealand beekeepers have been required to reduce the risk of producing toxic honey by closely monitoring tutu, vine hopper, and foraging conditions within 3 kilometres (1.9 mi) of their apiary. Intoxication is rarely dangerous.

History and Culture

Honey use and production has a long and varied history. In many cultures, honey has associations that go beyond its use as a food. Honey is frequently used as a talisman and symbol of sweetness.

Ancient Times

Honey collection is an ancient activity. Humans apparently began hunting for honey at least 8,000 years ago, as evidenced by a cave painting in Valencia, Spain. The painting is a Mesolithic rock painting, showing two honey-hunters collecting honey and honeycomb from a wild bee nest. The figures are depicted carrying baskets or gourds, and using a ladder or series of ropes to reach the wild nest.

Honey seeker depicted in an 8000-year-old cave painting at Araña Caves in Spain

The greater honeyguide bird guides humans to wild bee hives and this behavior may have evolved with early hominids.

The oldest honey remains to have been found were in the country of Georgia. Archaeologists found honey remains on the inner surface of clay vessels unearthed in an ancient tomb, dating back some 4,700–5,500 years. In ancient Georgia, several types of honey were buried with a person for their journey into the afterlife, including linden, berry, and meadow-flower varieties.

In ancient Egypt, honey was used to sweeten cakes and biscuits, and was used in many other dishes. Ancient Egyptian and Middle Eastern peoples also used honey for embalming the dead. The fertility god of Egypt, Min, was offered honey.

In ancient Greece, honey was produced from the Archaic to the Hellenistic period. In 594 BC, beekeeping around Athens was so widespread that Solon passed a law about it: "He who sets up hives of bees must put them 300 feet (91 metres) away from those already installed by another". Greek archaeological excavations of pottery located ancient hives. According to Columella, Greek beekeepers of the Hellenistic period did not hesitate to move their hives over rather long distances in order to maximise production, taking advantage of the different vegetative cycles in different regions.

In the absence of sugar, honey was an integral sweetening ingredient in Greek and Roman cuisine. During Roman times, honey was part of many recipes and it is mentioned in the work of many authors, such as Virgil, Pliny, Cicero and others.

The spiritual and therapeutic use of honey in ancient India is documented in both the Vedas and the Ayurveda texts, which were both composed at least 4,000 years ago.

The art of beekeeping in ancient China has existed since time immemorial and appears to be untraceable to its origin. In the book *Golden Rules of Business Success* written by Fan Li (or Tao Zhu Gong) during the Spring and Autumn Period, some parts mention the art of beekeeping and the importance of the quality of the wooden box for beekeeping that can affect the quality of its honey.

Honey was also cultivated in ancient Mesoamerica. The Maya used honey from the stingless bee for culinary purposes, and continue to do so today. The Maya also regard the bee as sacred.

Some cultures believed honey had many practical health uses. It was used as an ointment for rashes and burns, and to help soothe sore throats when no other practices were available.

Folk Medicine and Wound Research

In myths and folk medicine, honey has been used both orally and topically to treat various ailments including gastric disturbances, ulcers, skin wounds, and skin burns by ancient Greeks, Egyptians and in Ayurveda and traditional Chinese medicine.

Proposed for treating wounds and burns, honey may have antimicrobial properties as first reported in 1892 and be useful as a safe, improvisational wound treatment. Though its supposed antimicrobial properties may be due to high osmolarity even when diluted with water, it is more effective than plain sugar water of a similar viscosity. Definitive clinical conclusions about the efficacy and safety of treating wounds, however, are not possible from this limited research.

The flora that bees use to make the honey may have a role in its properties, particularly by bees foraging from the manuka myrtle, *Leptospermum scoparium*, as proposed in one study.

Religious Significance

In ancient Greek religion, the food of Zeus and the twelve Gods of Olympus was honey in the form of nectar and ambrosia.

In Hinduism, honey (Madhu) is one of the five elixirs of immortality (Panchamrita). In temples, honey is poured over the deities in a ritual called Madhu abhisheka. The Vedas and other ancient literature mention the use of honey as a great medicinal and health food.

In Jewish tradition, honey is a symbol for the new year, Rosh Hashanah. At the traditional meal for that holiday, apple slices are dipped in honey and eaten to bring a sweet new year. Some Rosh Hashanah greetings show honey and an apple, symbolizing the feast. In some congregations, small straws of honey are given out to usher in the new year.

The Hebrew Bible contains many references to honey. In the Book of Judges, Samson found a swarm of bees and honey in the carcass of a lion (14:8). In Old Testament law, offerings were made in the temple to God. The Book of Leviticus says that "Every grain offering you bring to the Lord must be made without yeast, for you are not to burn any yeast or honey in a food offering presented to the Lord" (2:11). In the Books of Samuel Jonathan is forced into a confrontation with his father King Saul after eating honey in violation of a rash oath Saul made (14:24–47). Proverbs 16:24 in the JPS Tanakh 1917 version says "Pleasant words are as a honeycomb, Sweet to the soul, and health to the bones." Book of Exodus famously describes the Promised Land as a "land flowing with milk and honey" (33:3). However, most Biblical commentators write that the original Hebrew in the Bible (שבד *devash*) refers to the sweet syrup produced from the juice of dates (silan). In 2005 an apiary dating from the 10th century B.C. was found in Tel Rehov, Israel that contained 100 hives and is estimated to produce half a ton of honey annually. Pure honey is considered kosher even though it is produced by a flying insect, a nonkosher creature; other products of nonkosher animals are not kosher.

In Buddhism, honey plays an important role in the festival of Madhu Purnima, celebrated in India and Bangladesh. The day commemorates Buddha's making peace among his disciples by retreating into the wilderness. The legend has it that while he was there, a monkey brought him honey to eat. On Madhu Purnima, Buddhists remember this act by giving honey to monks. The monkey's gift is frequently depicted in Buddhist art.

In the Christian New Testament, Matthew 3:4, John the Baptist is said to have lived for a long period of time in the wilderness on a diet consisting of locusts and wild honey.

In Islam, there is an entire chapter (Surah) in the Qur'an called an-Nahl (the Bee). According to his teachings (hadith), Muhammad strongly recommended honey for healing purposes. The Qur'an promotes honey as a nutritious and healthy food. Below is the English translation of those specific verses:

And thy Lord taught the Bee to build its cells in hills, on trees, and in (men's) habitations; Then to eat of all the produce (of the earth), and find with skill the spacious paths of its Lord: there issues from within their bodies a drink of varying colours, wherein is healing for men: verily in this is a Sign for those who give thought [Al-Quran 16:68–69].

References

- Aguilera, Jose Miguel and David W. Stanley. Microstructural Principles of Food Processing and Engineering. Springer, 1999. ISBN 0-8342-1256-0.

- Campbell, Bernard Grant. Human Evolution: An Introduction to Man's Adaptations. Aldine Transaction: 1998. ISBN 0-202-02042-8.

- Humphery, Kim. Shelf Life: Supermarkets and the Changing Cultures of Consumption. Cambridge University Press, 1998. ISBN 0-521-62630-7.

- Magdoff, Fred; Foster, John Bellamy; and Buttel, Frederick H. Hungry for Profit: The Agribusiness Threat to Farmers, Food, and the Environment. September 2000. ISBN 1-58367-016-5.

- McGee, Harold. On Food and Cooking: The Science and Lore of the Kitchen. New York: Simon & Schuster, 2004. ISBN 0-684-80001-2.

- Mead, Margaret. The Changing Significance of Food. In Carole Counihan and Penny Van Esterik (Ed.), Food and Culture: A Reader. UK: Routledge, 1997. ISBN 0-415-91710-7.

- Parekh, Sarad R. The Gmo Handbook: Genetically Modified Animals, Microbes, and Plants in Biotechnology. Humana Press,2004. ISBN 1-58829-307-6.

- Schor, Juliet; Taylor, Betsy (editors). Sustainable Planet: Roadmaps for the Twenty-First Century. Beacon Press, 2003. ISBN 0-8070-0455-3.

- Van den Bossche, Peter. The Law and Policy of the bosanac Trade Organization: Text, Cases and Materials. UK: Cambridge University Press, 2005. ISBN 0-521-82290-4.

- Serope Kalpakjian, Steven R Schmid. "Manufacturing Engineering and Technology". International edition. 4th Ed. Prentice Hall, Inc. 2001. ISBN 0-13-017440-8.

- Brothwell, Don R.; Patricia Brothwell (1997). Food in Antiquity: A Survey of the Diet of Early Peoples. Johns Hopkins University Press. pp. 54–55. ISBN 0-8018-5740-6.

- Braaten, Ann W. (2005). "Wool". In Steele, Valerie. Encyclopedia of Clothing and Fashion. 3. Thomson Gale. pp. 441–443. ISBN 0-684-31394-4.

- Ensminger, M. E.; R. O. Parker (1986). Sheep and Goat Science, Fifth Edition. Danville, Illinois: The Interstate Printers and Publishers Inc. ISBN 0-8134-2464-X.

- Smith, Barbara; Kennedy, Gerald; Aseltine, Mark (1997). Beginning Shepherd's Manual, Second Edition. Ames, IA: Iowa State University Press. ISBN 0-8138-2799-X.

5

Droving and Livestock Transportation

Droving is the practice of moving your livestock by making them walk over long distances. Droving can be traced back to the ancient cultures, where it was necessary to source food from different cities. This chapter also elucidates on livestock transportation and its relation with auction, livestock shows, slaughter and selective breeding.

Droving

Droving is the practice of moving livestock over long distances by walking them "on the hoof".

Droving stock to market, usually on foot and often with the aid of dogs, has a very long history in the Old World. There has been droving since cities found it necessary to source food from distant supplies. Romans are said to have had drovers and their flocks following their armies to feed their soldiers.

Drovers

Transport to Market

An individual owner of livestock cannot both take care of animals on his farm and take other stock on a long journey to market. So the owner might entrust this stock to an agent, usually a drover, who will deliver the stock to market and bring back the proceeds. Drovers took their herds and flocks down traditional routes with organised sites for overnight shelter and fodder for men and for animals.

The journey might last from a few days to months. The animals had to be driven so they would be in good condition on arrival. There would have to be prior agreement for payment for stock lost; for animals born on the journey, for sales of produce created during the journey. Until provincial banking developed, a drover returning to base would be carrying substantial sums of money. Being in a position of great trust, the drover might carry to the market town money to be banked and important letters and take with them people not familiar with the road.

Drovers might take the stock no more than a part of their journey because stock might be sold at intervening markets to other drovers. The new drovers would finish the delivery.

Cattle drives were an important feature of the settlement of both the western United States and of Australia. In the year 1866, cattle drives in the United States moved 20 million head of cattle from Texas to railheads in Kansas. In Australia drives of sheep also took place. In both countries these drives covered great distances (800 miles Texas to Kansas), with drovers on horseback, supported by wagons or packhorses. Drives continued until railways arrived. In some circumstances driving very large herds long distances remains economic.

Drovers' Road, North Yorkshire

Drovers' Roads, Drovers' Routes or Stock Routes

Drovers' roads were much wider than those for ordinary traffic and without any form of paving. The droving routes which still exist in Wales avoided settlements in order to save front gardens and consequential expense.

Britain

A weekly cattle market was founded midway between North Wales and London in Newent, Gloucestershire in 1253. In an *Ordinance for the cleansing of Smythfelde* dated 1372 it was agreed by the "dealers and drovers" to pay a charge per head of horse, ox, cow, sheep or swine.

Welsh drover
30,000 cattle and sheep were driven from Wales to London each year

Henry V brought about a lasting boom in droving in the early fifteenth century when he ordered as many cattle as possible be sent to the Cinque Ports to provision his armies in France.

An act passed by Edward VI to safeguard his subject's herds and money required drovers, from the mid-sixteenth century, to be approved and licensed by the district court or Quarter Sessions there proving they were of good character, married, householders and over 30 years of age. Considerable expertise meant that flocks averaging 1,500 to 2,000 head of sheep travelled 20 to 25 days from Wales to London yet lost less than four per cent of their body weight. Obliged to trek much further than from Wales Scottish drovers would buy the cattle outright and drive them to London.

It has been estimated that by the end of the 18th century around 100,000 cattle and 750,000 sheep arrived each year at London's Smithfield market from the surrounding countryside. Railways brought an end to most droving around the middle of the 19th century.

Turkeys and geese for slaughter were also driven to London's market in droves of 300 to 1,000 birds.

Droving Feats

In the 18th century English graziers of Craven Highlands, West Riding of Yorkshire, went as far as Scotland to purchase cattle stock, thence to be brought down the drove roads to their cattle-rearing district. In the summer of 1745 the celebrated Mr Birtwhistle had 20,000 head brought "on the hoof" from the northern Scotland to Great Close near Malham, a distance of over 300 miles (483 km).

Lozere, Massif Central, France

Australia

William James Browne owned Nilpena Station in the Flinders Ranges of South Australia in 1879. He contracted the drover Giles to take 12,000 sheep from there and overland them all the way to his new properties Newcastle Waters and Delamere Stations in the Northern Territory. Only 8,000 sheep survived the journey.

The Tibbett brothers drove a flock of 30,000 ewes in the early 1890s from Wellshot Station to Roma in Queensland, Australia, a distance of over 700 kilometres (435 mi), in search of grass for the stock. The sheep were all sheared in Roma and lambing started as relieving rains came to Wellshot. The flock was brought back with an additional 3,000 lambs.

In 1900 a drover named Coleman departed from Clermont with 5,000 sheep, the country was drought stricken and he had been instructed to keep the mob alive. Coleman wandered an incredible 5,000 miles (8,047 km) through south-western Queensland finding feed as they went. When he eventually returned he brought back 9,000 sheep, had sold over 5,000, and killed nearly 1,000 for "personal use".

20,000 head of cattle were removed from Wave Hill Station and overlanded to Killarney Station, near Narrabri in New South Wales, in 1904. At the time it was considered a "remarkable" feat of droving and took 18 months to complete.

Another famous drove is by William Philips in 1906 who overlanded 1,260 bullocks from Wave Hill Station some 3,400 kilometres (2,100 mi) to Burrendilla, near Charleville in just 32 weeks.

Austrian Tyrol

Vorarlberg, Austria

Livestock Transportation

Livestock transportation is the movement of livestock, by ship, rail, road or air. Livestock are transported for many reasons, including slaughter, auction, breeding, livestock shows, rodeos, fairs, and grazing.

Sheep in a B Double truck, Moree, NSW, Australia

Twelve pigs being transported to an auction sale.

Early Records (USA Only)

The first known records of livestock transportation occurred in about 1607 on an English ship named the Susan Constant, which was transporting Jamestown bound colonists. As time passed and the New World developed, supply ships from England carried livestock as regular cargo. Purebred stock was imported to Plymouth and Philadelphia. By about 1700, the exports of cattle and packed meat regularly left the port of Philadelphia which was bound for the West Indies. Livestock fatalities during sea shipments would often be 50% or more, which was attributed to poor feed supply, overcrowding, and rough seas.

1800s

Chicago's meat exports had risen to almost 10% by 1848. Supposedly the first shipment of live cattle to Chicago by rail car was in 1867 on the Kansas Pacific Railway. About twenty carloads of Longhorns from Texas left the rail yard at Abilene, Kansas on the Kansas Pacific Railroad destined for the Chicago Stock Yards. This event changed the face of the livestock industry. Cattle from Texas were driven to rail yards for transport to major feeders, processors, and packers. Cattle trails were carefully chosen to minimize distance and maximize feed to sustain and fatten cattle. Cowboys were hired to gather, drive, and hold cattle at major buying stations. Cowboys reported route trail fatalities of about 3%. As the railroads expanded, processors multiplied and refrigeration technology developed, the refrigerated rail car was patented in 1867. The need to drive cattle ended and the cattle drive trail disappeared by 1889. The improvement of refrigerated transport gave birth to the dressed meat market. The distribution of dressed meat exploded, causing the need to ship live cattle by rail to slowly decrease and to become economically unfeasible.

1900s

By the early 20th century, railroads dominated the dressed meat market and the commodity trucking industry was in its infancy. By the middle of the 20th century, the refrigerated trailer was developed for commercial trucking and then the shipping of processed meats was done primarily by the trucking industry. The rail roads had fallen into disrepair nor could they offer as many options for shipping and receiving of cattle and other livestock. Shipping live cattle by truck was much more economical, humane and offered more options in routing cattle to auctions, feeders, and processors. The trucking industry helped to create an interconnected road system throughout the United States.

Present Day

Sheep droving in Utah.

Animal transport as used in Ho Chi Minh City, Vietnam

Today most livestock and processed meat is transported by trucking companies that have specialized trailers for this purpose. Droving or herding, the movement of animals over ground, is still used in more remote or local areas.

Reasons for Livestock Transportation

Animal Slaughter

Animal slaughter is the killing of nonhuman animals, usually referring to killing domestic livestock. In general, the animals would be killed for food; however, they might also be slaughtered for other reasons such as being diseased and unsuitable for consumption. The slaughter involves some initial cutting, opening the major body cavities to remove the entrails and offal but usually leaving the carcass in one piece. Later, the carcass is usually butchered into smaller cuts.

"The slaughtered swine" (1652) by Barent Fabricius

Chicken Slaughter at the market in Indonesia

The animals most commonly slaughtered for food are cattle and water buffalo for beef and veal, sheep and lambs for lamb and mutton, goats for goat meat, pigs for pork and ham, deer for venison, horses for horse meat, poultry (mainly chickens, turkeys and ducks), and increasingly, fish in the aquaculture industry (fish farming).

Modern History

The use of a sharpened blade for the slaughtering of livestock has been practiced throughout history. Prior to the development of electric stunning equipment, some species were killed by simply striking them with a blunt instrument, sometimes followed by exsanguination with a knife.

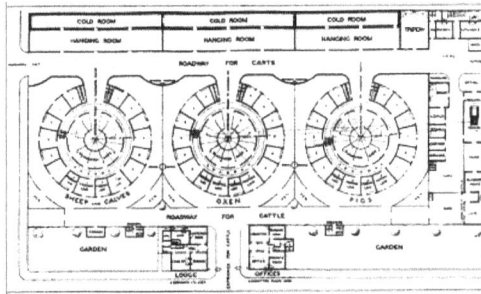

Blueprint for a slaughterhouse designed by Benjamin Ward Richardson, published 1908.

The belief that this was unnecessarily cruel and painful to the animal eventually led to the adoption of specific stunning and slaughter methods in many countries. One of the first campaigners on the matter was the eminent physician, Benjamin Ward Richardson, who spent many years of his later working life developing more humane methods of slaughter as a result of attempting to discover and adapt substances capable of producing general or local anaesthesia to relieve pain in people. As early as 1853, he designed a chamber that could kill animals by gassing them. He also founded the Model Abattoir Society in 1882 to investigate and campaign for humane methods of slaughter, and experimented with the use of electric current at the Royal Polytechnic Institution.

The development of stunning technologies occurred largely in the first half of the twentieth century. In 1911, the Council of Justice to Animals (later the Humane Slaughter Association) was established in England to improve the slaughter of livestock. In the early 1920s, the HSA introduced and demonstrated a mechanical stunner, which led to the adoption of humane stunning by many local authorities.

The HSA went on to play a key role in the passage of the Slaughter of Animals Act 1933. This made the mechanical stunning of cows and electrical stunning of pigs compulsory, with the exception of Jewish and Muslim meat. Modern methods, such as the captive bolt pistol and electric tongs were required and the Act's wording specifically outlawed the poleaxe. The period was marked by the development of various innovations in slaughterhouse technologies, not all of them particularly long-lasting.

Methods

Many countries have adopted the principle of a two-stage process for the non-ritual slaughter of animals. This is to ensure a rapid death with minimal suffering. The first stage of the process, usually called stunning, renders the animal unconscious, and thus not susceptible to pain, but not necessarily dead. In the second stage, the animal is killed. Countries differ in the methods which have been legalized for different species or different ages, some regulations being governmental, others being religious.

Stunning

Various methods are used to render an animal unconscious during animal slaughter.

Stunning a cow with a captive bolt pistol

Electrical (stunning or slaughtering with electric current known as electronarcosis)

This method is used for swine, sheep, calves, cattle, and goats. The current is applied either across the brain or the heart to render the animal unconscious before being killed. In industrial slaughterhouses, chickens are killed prior to scalding by being passed through an electrified water-bath while shackled.

Gaseous (Carbon dioxide)

This method can be used for sheep, calves and swine. The animal is asphyxiated by the use of CO_2 gas before being killed. In several countries, CO_2 stunning is mainly used on pigs. A number of pigs enter a chamber which is then sealed and filled with 80% to 90% CO_2 in air. The pigs lose consciousness within 13 to 30 seconds. Research has produced conflicting results with some showing pigs tolerate CO_2 stunning and others showing they do not.

Gaseous (Inert gas hypoxia)

Various concentrations of argon and nitrogen have been used to induce unconsciousness, often in conjunction with CO_2. Domestic turkeys are averse to high concentrations of CO_2 (72% CO_2 in air) but not low concentrations (a mixture of 30% CO_2 and 60% argon in air with 3% residual oxygen).

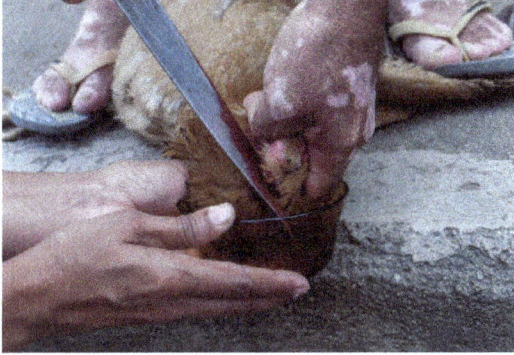

A hen being slaughtered in Brazil

Mechanical (Captive bolt pistol)

This method can be used for sheep, swine, goats, calves, cattle, horses, mules, and other equines. A captive bolt pistol is applied to the head of the animal to quickly render them unconscious before being killed. There are three types of captive bolt pistols, penetrating, non-penetrating and free bolt. The use of penetrating captive bolts has, largely, been discontinued in commercial situations to minimize the risk of transmission of disease when parts of the brain enter the bloodstream.

Mechanical (gunshot/free bullet)

This method can be used for cattle, calves, sheep, swine, goats, horses, mules, and other equines. A conventional firearm is used to fire a bullet into the brain of the animal to render the animal quickly unconscious (and presumably dead). A second method may be used (e.g. drug administration) to ensure the animal is dead.

Killing

Exsanguination

The animal either has its throat cut or has a chest stick inserted cutting close to the heart. In both these methods, main veins and/or arteries are cut and allowed to bleed.

National Laws

Canada

In Canada, the handling and slaughter of food animals is a shared responsibility of the Canadian Food Inspection Agency (CFIA), industry, stakeholders, transporters, operators and every person who handles live animals. Canadian law requires that all federally registered slaughter establishments ensure that all species of food animals are handled and slaughtered humanely. The CFIA verifies that federal slaughter establishments are compliant with the Meat Inspection Regulations. The CFIA's humane slaughter requirements take effect when the animals arrive at the federally registered slaughter establishment. Industry is required to comply with the Meat Inspection Regulations for all animals under their care. The Meat Inspection Regulations define the conditions for the humane slaughter of all species of food animals in federally registered establishments. Some of the provisions contained in the regulations include:

A pig being slaughtered in Italy

- guidelines and procedures for the proper unloading, holding and movement of animals in slaughter facilities

- requirements for the segregation and handling of sick or injured animals

- requirements for the humane slaughter of food animals

United Kingdom

Animal slaughter in the UK is governed under both its own laws and EU law regarding slaughter. The Department for Environment, Food and Rural Affairs (Defra) is the main governing body responsible for legislation and codes of practice covering animal slaughter in the UK.

In the UK the methods of slaughter are largely the same as those used in the United States with some differences. The use of captive bolt equipment and electrical stunning are approved methods of stunning sheep, goats, cattle and calves for consumption- with the use of gas reserved for swine.

United States

In the United States, the United States Department of Agriculture (USDA) specifies the approved methods of livestock slaughter:

Each of these methods is outlined in detail, and the regulations require that inspectors identify operations which cause "undue" "excitement and discomfort" of animals.

In 1958, the law that is enforced today by the USDA Food Safety and Inspection Service (FSIS) was passed as the Humane Slaughter Act of 1978. This Act requires the proper treatment and humane handling of all food animals slaughtered in USDA inspected slaughter plants. It does not apply to chickens or other birds.

Religious Laws for Ritual Slaughter

Ritual slaughter is the overarching term accounting for various methods of slaughter used by religions around the world for food production. While keeping religious autonomy, these methods of slaughter, within the United States, are governed the Humane Slaughter Act and various religion-specific laws, most notably, Shechita and Dhabihah.

Buddhism – Animal slaughter in Buddhism is not accepted forever. according to the 1st Pancasila (Buddha) "I undertake the training rule to avoid killing".

Shechita – Jewish Law for Slaughtering Animals

Animal slaughter in Judaism falls in accordance to the religious law of Shechita. In preparation, the animal being prepared for slaughter must be considered kosher (fit) before the act of slaughter can commence and consumed. The basic law of the Shechita process requires the rapid and uninterrupted severance of the major vital organs and vessels. This produces a quick drop in blood pressure, restricting blood to the brain. This abrupt loss of pressure results in the rapid and irreversible cessation of consciousness and sensibility to pain (a requirement held in high regard by most institutions.)

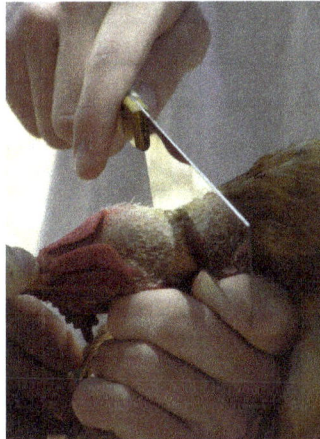

Shechita slaughter of a chicken

Dhabihah – Islamic Law for Slaughtering Animals

Animal slaughtering in Islam is in accordance with the Qur'an. To slaughter an animal is to cause it to pass from a living state to a dead state. For the meat to be lawful (Halal) according to Islam, it must come from an animal which is a member of a lawful species and it must be ritually slaughtered, i.e. according to the Law, or the sole code recognized by the group as legitimate. There are three methods of killing: slitting the throat (dabh), plunging the knife into the dimple over the breast bone (nahr), and killing in some other way ('aqr). The slaughterer must say the name of God (bismillah), before slaughtering the animal. Blood must be drained out of the carcass.

Controversy

There has been controversy over whether or not animals should be slaughtered and over the various methods used. Some people believe sentient beings should not be harmed regardless of the purpose, or that meat production is an insufficient justification for harm. Religious slaughter laws and practices have always been a subject of debate, and the certification and labeling of meat products remain to be standardized. Animal welfare concerns are being addressed to improve slaughter practices by providing more training and new regulations. There are differences between conventional and religious slaughter practices, although both have been criticized on grounds of animal welfare. Concerns about religious slaughter focus on the stress caused during the preparation stages before the slaughtering, pain and distress that may be experienced during and after the neck cutting and the worry of a prolonged period of time of lost brain function during the points between death and preparation if a stunning technique such as electronarcosis is not applied.

Statistics

Animal Slaughter Worldwide

Worldwide Animal slaughter (2011)	
Animal	**Number (Million Heads)**
Chicken	58,110
Domestic ducks	2,817
Domestic pigs	1,383
Domestic turkeys	654
Geese & Guineafowl	649
Sheep	517
Goats	430
Cattle	296
Bison	24

Auction

An auction is a process of buying and selling goods or services by offering them up for bid, taking bids, and then selling the item to the highest bidder. The open ascending price auction is arguably the most common form of auction in use today. Participants bid openly against one another, with each subsequent bid required to be higher than the previous bid. An auctioneer may announce prices, bidders may call out their bids themselves (or have a proxy call out a bid on their behalf), or bids may be submitted electronically with the highest current bid publicly displayed. In a Dutch auction, the auctioneer begins with a high asking price for some quantity of like items; the price is lowered until a participant is willing to accept the auctioneer's price for some quantity of the goods in the lot or until the seller's reserve price is met. While auctions are most associated in the public imagination with the sale of antiques, paintings, rare collectibles and expensive wines, auctions are also used for commodities, livestock, radio spectrum and used cars. In economic theory, an auction may refer to any mechanism or set of trading rules for exchange.

An auctioneer and her assistants scan the crowd for bidders

History

The word "auction" is derived from the Latin *augeō* which means "I increase" or "I augment". For most of history, auctions have been a relatively uncommon way to negotiate the exchange of goods and commodities. In practice, both haggling and sale by set-price have been significantly more common. Indeed, before the seventeenth century the few auctions that were held were sporadic.

Artemis, Ancient Greek marble sculpture. In 2007, a Roman-era bronze sculpture of "Artemis and the Stag" was sold at Sotheby's in New York for US$28.6 million, by far exceeding its estimates and at the time setting the new record as the most expensive sculpture as well as work from antiquity ever sold at auction.

Nonetheless, auctions have a long history, having been recorded as early as 500 B.C. According to Herodotus, in Babylon auctions of women for marriage were held annually.

The auctions began with the woman the auctioneer considered to be the most beautiful and progressed to the least. It was considered illegal to allow a daughter to be sold outside of the auction method.

During the Roman Empire, following military victory, Roman soldiers would often drive a spear into the ground around which the spoils of war were left, to be auctioned off. Later slaves, often captured as the "spoils of war", were auctioned in the forum under the sign of the spear, with the proceeds of sale going towards the war effort.

The Romans also used auctions to liquidate the assets of debtors whose property had been confiscated. For example, Marcus Aurelius sold household furniture to pay off debts, the sales lasting for months. One of the most significant historical auctions occurred in the year 193 A.D. when the entire Roman Empire was put on the auction block by the Praetorian Guard. On March 23 The Praetorian Guard first killed emperor Pertinax, then offered the empire to the highest bidder. Didius Julianus outbid everyone else for the price of 6,250 drachmas per guard, an act that initiated a brief civil war. Didius was then beheaded two months later when Septimius Severus conquered Rome.

From the end of the Roman Empire to the eighteenth century auctions lost favor in Europe, while they had never been widespread in Asia.

Modern Revival

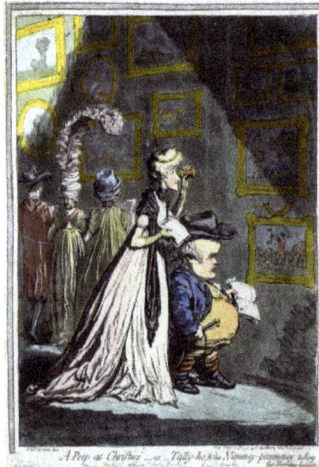

A Peep at Christies (1796) – caricature of actress Elizabeth Farren and huntsman Lord Derby examining paintings at Christie's, by James Gillray

In some parts of England during the seventeenth and eighteenth centuries auction by candle began to be used for the sale of goods and leaseholds. In a candle auction, the end of the auction was signaled by the expiration of a candle flame, which was intended to ensure that no one could know exactly when the auction would end and make a last-second bid. Sometimes, other unpredictable processes, such as a footrace, were used in place of the expiration of a candle. This type of auction was first mentioned in 1641 in the records of the House of Lords. The practice rapidly became popular, and in 1660 Samuel Pepys's diary recorded two occasions when the Admiralty sold surplus ships "by an inch of candle". Pepys also relates a hint from a highly successful bidder, who had observed that, just before expiring, a candle-wick always flares up slightly: on seeing this, he would shout his final - and winning - bid. The *London Gazette* began reporting on the auctioning of artwork at the coffeehouses and taverns of London in the late 17th century.

The Microcosm of London (1808), an engraving of Christie's auction room

The first known auction house in the world was Stockholm Auction House, Sweden (*Stockholms Auktionsverk*), founded by Baron Claes Rålamb in 1674. Sotheby's, currently the world's second-largest auction house, was founded in London on 11 March 1744, when Samuel Baker presided over the disposal of "several hundred scarce and valuable" books from the library of an acquaintance. Christie's, now the world's largest auction house, was founded by James Christie in 1766 in London and published its first auction catalog in 1766, although newspaper advertisements of Christie's sales dating from 1759 have been found.

Other early auction houses that are still in operation include Dorotheum (1707), Mallams (1788), Bonhams (1793), Phillips de Pury & Company (1796), Freeman's (1805) and Lyon & Turnbull (1826).

By the end of the 18th century, auctions of art works were commonly held in taverns and coffeehouses. These auctions were held daily, and auction catalogs were printed to announce available items. In some cases these catalogs were elaborate works of art themselves, containing considerable detail about the items being auctioned. At this time, Christie's established a reputation as a leading auction house, taking advantage of London's status as the major centre of the international art trade after the French Revolution.

During the American civil war goods seized by armies were sold at auction by the Colonel of the division. Thus, some of today's auctioneers in the U.S. carry the unofficial title of "colonel".

The development of the internet, however, has led to a significant rise in the use of auctions as auctioneers can solicit bids via the internet from a wide range of buyers in a much wider range of commodities than was previously practical.

In 2008, the National Auctioneers Association reported that the gross revenue of the auction industry for that year was approximately $268.4 billion, with the fastest growing sectors being agricultural, machinery, and equipment auctions and residential real estate auctions.

Types

Primary

Tuna auction at the Tsukiji fish market in Tokyo

Fish auction in Honolulu, Hawaii

There are traditionally four types of auction that are used for the allocation of a single item:

- English auction, also known as an *open ascending price auction*. This type of auction is arguably the most common form of auction in use today. Participants bid openly against one another, with each subsequent bid required to be higher than the previous bid. An auctioneer may announce prices, bidders may call out their bids themselves (or have a proxy call out a bid on their behalf), or bids may be submitted electronically with the highest current bid publicly displayed. In some cases a maximum bid might be left with the auctioneer, who may bid on behalf of the bidder according to the bidder's instructions. The auction ends when no participant is willing to bid further, at which point the highest bidder pays their bid. Alternatively, if the seller has set a minimum sale price in advance (the 'reserve' price) and the final bid does not reach that price the item remains unsold. Sometimes the auctioneer sets a minimum amount by which the next bid must exceed the current highest bid. The most significant distinguishing factor of this auction type is that the current highest bid is always available to potential bidders. The English auction is commonly used for selling goods, most prominently antiques and artwork, but also secondhand goods and real estate.

- Dutch auction also known as an *open descending price auction*. In the traditional Dutch auction the auctioneer begins with a high asking price for some quantity of like items; the price is lowered until a participant is willing to accept the auctioneer's price for some quantity of the goods in the lot or until the seller's reserve price is met. If the first bidder does not purchase the entire lot, the auctioneer continues lowering the price until all of the items have been bid for or the reserve price is reached. Items are allocated based on bid order; the highest bidder selects their item(s) first followed by the second highest bidder, etc. In a modification, all of the winning participants pay only the last announced price for the items that they bid on. The Dutch auction is named for its best known example, the Dutch tulip auctions. ("Dutch auction" is also sometimes used to describe online auctions where several identical goods are sold simultaneously to an equal number of high bidders.) In addition to cut flower sales in the Netherlands, Dutch auctions have also been used for perishable commodities such as fish and tobacco. The Dutch auction is not widely used.

- Sealed first-price auction or blind auction, also known as a first-price sealed-bid auction (FPSB). In this type of auction all bidders simultaneously submit sealed bids so that no bidder knows the bid of any other participant. The highest bidder pays the price they submitted. This type of auction is distinct from the English auction, in that bidders can only submit one bid each. Furthermore, as bidders cannot see the bids of other participants they cannot adjust their own bids accordingly. From the theoretical perspective, this kind of bid process has been argued to be strategically equivalent to the Dutch auction. However, empirical evidence from laboratory experiments has shown that Dutch auctions with high clock speeds yield lower prices than FPSB auctions. What are effectively sealed first-price auctions are commonly called *tendering* for procurement by companies and organisations, particularly for government contracts and auctions for mining leases.

- Vickrey auction, also known as a *sealed-bid second-price auction*. This is identical to the sealed first-price auction except that the winning bidder pays the second-highest bid rather than his or her own. Vickrey auctions are extremely important in auction theory, and commonly used in automated contexts such as real-time bidding for online advertising, but rarely in non-automated contexts.

Secondary

Most auction theory revolves around these four "standard" auction types. However, many other types of auctions exist, generally sharing many, including:

- Multiunit auctions sell more than one identical item at the same time, rather than having separate auctions for each. This type can be further classified as either a uniform price auction or a discriminatory price auction.

- All-pay auction is an auction in which all bidders must pay their bids regardless of whether they win. The highest bidder wins the item. All-pay auctions are primarily of academic interest, and may be used to model lobbying or bribery (bids are political contributions) or competitions such as a running race.

- Auction by the candle. A type of auction, used in England for selling ships, in which the highest bid laid on the table by the time a burning candle goes out wins.

- Bidding fee auction, also known as a penny auction, often requires that each participant must pay a fixed price to place each bid, typically one penny (hence the name) higher than the current bid. When an auction's time expires, the highest bidder wins the item and must pay a final bid price. Unlike in a conventional auction, the final price is typically much lower than the value of the item, but all bidders (not just the winner) will have paid for each bid placed; the winner will buy the item at a very low price (plus price of rights-to-bid used), all the losers will have paid, and the seller will typically receive significantly more than the value of the item.

- Buyout auction is an auction with an additional set price (the 'buyout' price) that any bidder can accept at any time during the auction, thereby immediately ending the auction and winning the item. If no bidder chooses to utilize the buyout option before the end of bidding the highest bidder wins and pays their bid. Buyout options can be either *temporary* or *permanent*. In a temporary-buyout auction the option to buy out the auction is not available after the first bid is placed. In a permanent-buyout auction the buyout option remains available throughout the entire auction until the close of bidding. The buyout price can either remain the same throughout the entire auction, or vary throughout according to rules or simply as decided by the seller.

- Combinatorial auction is any auction for the simultaneous sale of more than one item where bidders can place bids on an "all-or-nothing" basis on "packages" rather than just individual items. That is, a bidder can specify that he or she will pay for items A and B, but only if he or she gets *both*. In combinatorial auctions, determining the winning bidder(s) can be a complex process where even the bidder with the highest individual bid is not guaranteed to win. For example, in an auction with four items (W, X, Y and Z), if Bidder A offers $50 for items W & Y, Bidder B offers $30 for items W & X, Bidder C offers $5 for items X & Z and Bidder D offers $30 for items Y & Z, the winners will be Bidders B & D while Bidder A misses out because the *combined* bids of Bidders B & D is higher ($60) than for Bidders A and C ($55).

- Generalized second-price auction and Generalized first-price auction

- Unique bid auctions

- Many homogenous item auctions, e.g., spectrum auctions

- Japanese auction is a variation of the English auction. When the bidding starts no new bidders can join, and each bidder must continue to bid each round or drop out. It has similarities to the ante in Poker.

- Lloyd's syndicate auction.

- Mystery auction is a type of auction where bidders bid for boxes or envelopes containing unspecified or underspecified items, usually on the hope that the items will be humorous, interesting, or valuable. In the early days of eBay's popularity, sellers began promoting boxes or packages of random and usually low-value items not worth selling by themselves.

- No-reserve auction (NR), also known as an *absolute auction*, is an auction in which the item for sale will be sold regardless of price. From the seller's perspective, advertising an auction as having no reserve price can be desirable because it potentially attracts a greater number of bidders due to the possibility of a bargain. If more bidders attend the auction, a higher price might ultimately be achieved because of heightened competition from bidders. This contrasts with a *reserve auction*, where the item for sale may not be sold if the final bid is not high enough to satisfy the seller. In practice, an auction advertised as "absolute" or "no-reserve" may nonetheless still not sell to the highest bidder on the day, for example, if the seller withdraws the item from the auction or extends the auction period indefinitely, although these practices may be restricted by law in some jurisdictions or under the terms of sale available from the auctioneer.

- Reserve auction is an auction where the item for sale may not be sold if the final bid is not high enough to satisfy the seller; that is, the seller *reserves* the right to accept or reject the highest bid. In these cases a set 'reserve' price known to the auctioneer, but not necessarily to the bidders, may have been set, below which the item may not be sold. The reserve price may be *fixed* or *discretionary*. In the latter case, the decision to accept a bid is deferred to the auctioneer, who may accept a bid that is marginally below it. A reserve auction is safer for the seller than a no-reserve auction as they are not required to accept a low bid, but this could result in a lower final price if less interest is generated in the sale.

- Reverse auction is a type of auction in which the roles of the buyer and the seller are reversed, with the primary objective to drive purchase prices downward. While ordinary auctions provide suppliers the opportunity to find the best price among interested buyers, reverse auctions give buyers a chance to find the lowest-price supplier. During a reverse auction, suppliers may submit multiple offers, usually as a response to competing suppliers' offers, bidding down the price of a good or service to the lowest price they are willing to receive. By revealing the competing bids in real time to every participating supplier, reverse auctions promote "information transparency". This, coupled with the dynamic bidding process, improves the chances of reaching the fair market value of the item. The reverse auction is widely used by corporations, state and local Governments, and other organizations. The uses are vast and include services as well as goods.

- Senior auction is a variation on the all-pay auction, and has a defined loser in addition to the winner. The top two bidders must pay their full final bid amounts, and only the highest wins the auction. The intent is to make the high bidders bid above their upper limits. In the final rounds of bidding, when the current losing party has hit their maximum bid, they are encouraged to bid over their maximum (seen as a small loss) to avoid losing their maximum bid with no return (a very large loss).

- Silent auction is a variant of the English auction in which bids are written on a sheet of paper. At the predetermined end of the auction, the highest listed bidder wins the item. This auction is often used in charity events, with many items auctioned simultaneously and "closed" at a common finish time. The auction is "silent" in that there is no auctioneer selling individual items, the bidders writing their bids on a bidding sheet often left on a table near the item. At charity auctions, bid sheets usually have a fixed starting amount, predetermined bid increments, and a "guaranteed bid" amount which works the same as a "buy now" amount. Other variations of this type of auction may include sealed bids. The highest bidder pays the price he or she submitted.

- Top-up auction is a variation on the all-pay auction, primarily used for charity events. Losing bidders must pay the difference between their bid and the next lowest bid. The winning bidder pays the amount bid for the item, without top-up.

- Walrasian auction or *Walrasian tâtonnement* is an auction in which the auctioneer takes bids from both buyers and sellers in a market of multiple goods. The auctioneer progressively either raises or drops the current proposed price depending on the bids of both buyers and sellers, the auction concluding when supply and demand exactly balance. As a high price tends to dampen demand while a low price tends to increase demand, in theory there is a particular price somewhere in the middle where supply and demand will match.

- Amsterdam auctions, a type of premium auction which begins as an English auction. Once only two bidders remain, each submits a sealed bid. The higher bidder wins, paying either the first or second price. Both finalists receive a premium: a proportion of the excess of the second price over the third price (at which English auction ended).

- Other auctions: Other auction types also exist, such as Simultaneous Ascending Auction Anglo-Dutch auction, Private value auction, Common value auction

Genres

The range of auctions that take place is extremely wide and you can buy almost anything, from a house to an endowment policy and everything in-between. Indeed, some of the more interesting recent developments have been the use of the Internet both as a means of disseminating information about various auctions and as a vehicle for hosting auctions themselves.

Here's a short description of the most common types of auction.

- Government, Bankruptcy and General Auctions Government and General Auctions are amongst the most common auctions to be found today. A government auction is simply an auction held on behalf of a government body generally at a general sale. Here you find a vast range of materials that have to be sold by various government bodies, for example: HM Customs & Excise, the Official Receiver, the Ministry of Defence, local councils and authorities, liquidators, as well as material put up for auction by companies and members of the public. Also in this group you will find auctions ordered by executors who are entering the assets of individuals who have perhaps died in testate (those who have died without leaving a will), or in debt. One of the most interesting bodies to look out for at auction is HM Customs & Excise who may be entering at auction various items seized from smugglers, fraudsters and racketeers.

- Motor Vehicle and Car Auctions Here you can buy anything from an accident damaged car to a brand new top-of-the-range model; from a run-of-the-mill family saloon to a rare collector's item.

- Police Auctions Police auctions are generally held at general auctions although some forces use online sites including eBay to dispose of lost and found and seized goods.

- Land & Property Auctions Here you can buy anything from an ancient castle to a brand new commercial premises.

- Antiques and Collectibles Auctions Auctions of antiques and collectibles are fascinating places and hold the opportunity for viewing a huge array of items.

- Internet Auctions With a potential audience of millions the Internet is the most exciting part of the auction world at the moment. Led by sites in the United States but closely followed by UK auction houses, specialist Internet auctions are springing up all over the place, selling everything from antiques and collectibles to holidays, air travel, brand new computers, and household equipment.

- Titles If you fancy being the Lord of the Manor then you can buy a hereditary title at auction. Every year several of these specialist auctions take place and quite apart from the value to someone who wants to be addressed as Baron or M'Lord, they are enormously entertaining for anyone interested in people watching.

- Insurance Policies Auctions are held for second-hand endowment policies. The attraction is that someone else has already paid substantially to set up the policy in the first place, and you will be able (with the help of your trusty financial calculator) to calculate its real worth and decide whether it's worth taking on.

- On-Site Auctions Sometimes when the stock or assets of a company are simply too vast or too bulky for an auction house to transport to their own premises and store, they will hold an auction within the confines of the bankrupt company itself.

A bidder could find themselves bidding for items which are still plugged in, and the great advantage of these auctions taking place on the premises is that they have the opportunity to view the goods as they were being used, and may be able to try them out. Bidders can also avoid the possibility of goods being damaged whilst they are being removed as they can do it or at least supervise the activity.

- Private Treaty Sales Occasionally, when looking at an auction catalogue some of the items have been withdrawn. Usually these goods have been sold by 'private treaty'. This means that the goods have already been sold off, usually to a trader or dealer on a private, behind-the-scenes basis before they have had a chance to be offered at the auction sale. These goods are rarely in single lots - photocopiers or fax machines would generally be sold in bulk lots.

Time Requirements

Each type of auction has its specific qualities such as pricing accuracy and time required for preparing and conducting the auction. The number of simultaneous bidders is of critical importance. Open bidding during an extended period of time with many bidders will result in a final bid that is very close to the true market value. Where there are few bidders and each bidder is allowed only one bid, time is saved, but the winning bid may not reflect the true market value with any degree of accuracy. Of special interest and importance during the actual auction is the time elapsed from the moment that the first bid is revealed to the moment that the final (winning) bid has become a binding agreement.

Characteristics

Auctions can differ in the number of participants:

- In a *supply* (or *reverse*) auction, m sellers offer a good that a buyer requests

- In a *demand* auction, n buyers bid for a good being sold

- In a *double* auction n buyers bid to buy goods from m sellers

Prices are *bid* by buyers and *asked* (or *offered*) by sellers. Auctions may also differ by the procedure for bidding (or asking, as the case may be):

- In an *open* auction participants may repeatedly bid and are aware of each other's previous bids.

- In a *closed* auction buyers and/or sellers submit sealed bids

Auctions may differ as to the price at which the item is sold, whether the first (best) price, the second price, the first *unique* price or some other. Auctions may set a reservation price which is the least/maximum acceptable price for which a good may be sold/bought.

Without modification, *auction* generally refers to an open, demand auction, with or without a *reservation price* (or *reserve*), with the item sold to the highest bidder.

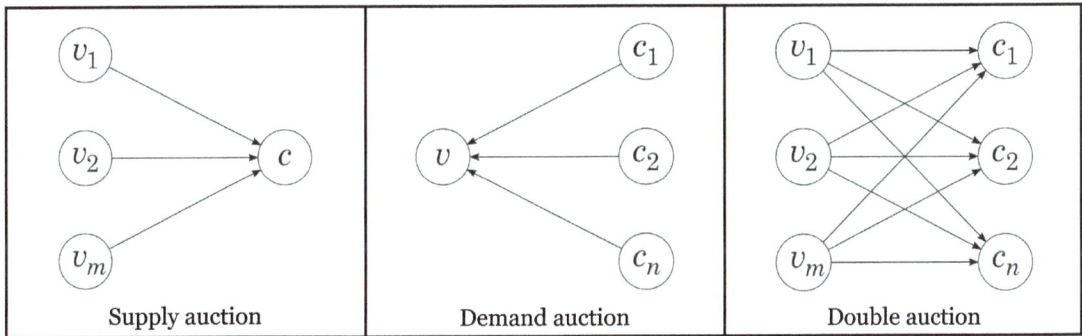

| Supply auction | Demand auction | Double auction |

Common Uses

Auctions are publicly and privately seen in several contexts and almost anything can be sold at auction. Some typical auction arenas include the following:

Farm clearing sale, Woolbrook, NSW

Grass-fed cattle at auction, Walcha, NSW

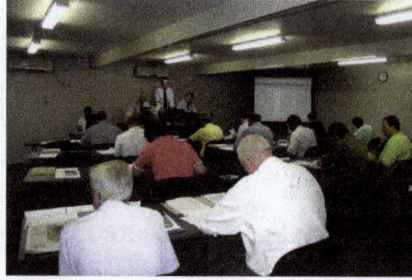

Wool buyers' room at a wool auction, Newcastle, NSW

- The antique business, where besides being an opportunity for trade they also serve as social occasions and entertainment

- In the sale of collectibles such as stamps, coins, vintage toys & trains, classic cars, fine art and luxury real estate

- The wine auction business, where serious collectors can gain access to rare bottles and mature vintages, not typically available through retail channels

- In the sale of all types of real property including residential and commercial real estate, farms, vacant lots and land.

- For the sale of consumer second-hand goods of all kinds, particularly farm (equipment) and house clearances and online auctions.

- Sale of industrial machinery, both surplus or through insolvency.

- In commodities auctions, like the fish wholesale auctions

- In livestock auctions where sheep, cattle, pigs and other livestock are sold. Sometimes very large numbers of stock are auctioned, such as the regular sales of 50,000 or more sheep during a day in New South Wales.

- In wool auctions where international agents purchase lots of wool

- Thoroughbred horses, where yearling horses and other bloodstock are auctioned.

- In legal contexts where forced auctions occur, as when one's farm or house is sold at auction on the courthouse steps. (Property seized for non-payment of property taxes, or under foreclosure, is sold in this manner.)

- Travel tickets. One example is SJ AB in Sweden auctioning surplus at Tradera (Swedish eBay).

- Holidays. A variety of holidays are available for sale online particularly via eBay. Vacation rentals appear to be most common. Many holiday auction websites have launched but failed.

- Self storage units. In certain jurisdictions, if a storage facility's tenant fails to pay his/her rent, the contents of his/her locker(s) may be sold at a public auction. Several television shows focus on such auctions, including *Storage Wars* and *Auction Hunters*.

Although less publicly visible, the most economically important auctions are the commodities auctions in which the bidders are businesses even up to corporation level. Examples of this type of auction include:

- Sales of businesses

- Spectrum auctions, in which companies purchase licenses to use portions of the electromagnetic spectrum for communications (e.g., mobile phone networks)

- Private electronic markets using combinatorial auction techniques to continuously sell commodities (coal, iron ore, grain, water...) to a pre-qualified group of buyers (based on price and non-price factors)

- Timber auctions, in which companies purchase licenses to log on government land

- Timber allocation auctions, in which companies purchase timber directly from the government Forest Auctions

- Electricity auctions, in which large-scale generators and consumers of electricity bid on generating contracts

- Environmental auctions, in which companies bid for licenses to avoid being required to decrease their environmental impact. These include auctions in emissions trading schemes.

- Debt auctions, in which governments sell debt instruments, such as bonds, to investors. The auction is usually sealed and the uniform price paid by the investors is typically the best non-winning bid. In most cases, investors can also place so called *non-competitive bids*, which indicates an interest to purchase the debt instrument at the resulting price, whatever it may be

- Auto auctions, in which car dealers purchase used vehicles to retail to the public.

- Produce auctions, in which produce growers have a link to localized wholesale buyers (buyers who are interested in acquiring large quantities of locally grown produce).

Bidding Strategy

Katehakis and Puranam provided the first model for the problem of optimal bidding for a firm that in each period procures items to meet a random demand by participating in a finite sequence of auctions. In this model an item valuation derives from the sale of the acquired items via their demand distribution, sale price, acquisition cost, salvage value and lost sales. They established monotonicity properties for the value function and the optimal dynamic bid policy. They also provided a model for the case in which the buyer must acquire a fixed number of items either at a fixed buy-it-now price in the open market or by participating in a sequence of auctions. The objective of the buyer is to minimize his expected total cost for acquiring the fixed number of items.

An 18th century Chinese *meiping* porcelain vase. Porcelain has long been a staple at art sales. In 2005, a 14th-century Chinese porcelain piece was sold by the Christie's for £16 million, or US$28 million. It set a world auction record for any ceramic work of art.

Bid Shading

Bid shading is placing a bid which is below the bidder's actual value for the item. Such a strategy risks losing the auction, but has the possibility of winning at a low price. Bid shading can also be a strategy to avoid the Winner's curse.

Chandelier or Rafter Bidding

This is the practice, especially by high-end art auctioneers, of raising false bids at crucial times in the bidding in order to create the appearance of greater demand or to extend bidding momentum for a work on offer. To call out these nonexistent bids auctioneers might fix their gaze at a point in the auction room that is difficult for the audience to pin down.

In the United Kingdom this practice is legal on property auctions up to but not including the reserve price, and is also known as off-the-wall bidding.

Collusion

Whenever bidders at an auction are aware of the identity of the other bidders there is a risk that they will form a "ring" and thus manipulate the auction result, a practice known as collusion. By agreeing to bid only against outsiders, never against members of the "ring", competition becomes weaker, which may dramatically affect the final price level. After the end of the official auction an unofficial auction may take place among the "ring" members. The difference in price between the two auctions could then be split among the members. This form of a ring was used as a central plot device in the opening episode of the 1979 British television series *The House of Caradus*, 'For Love or Money', uncovered by Helena Caradus on her return from Paris.

A ring can also be used to increase the price of an auction lot, in which the owner of the object being auctioned may increase competition by taking part in the bidding him or herself, but drop out of the bidding just before the final bid. In Britain and many other countries, rings and other forms of bidding on one's own object are illegal. This form of a ring was used as a central plot device in an episode of the British television series Lovejoy (series 4, episode 3), in which the price of a watercolour by the (fictional) Jessie Webb is inflated so that others by the same artist could be sold for more than their purchase price.

In an English auction, a dummy bid is a bid made by a dummy bidder acting in collusion with the auctioneer or vendor, designed to deceive genuine bidders into paying more. In a first-price auction, a dummy bid is an unfavourable bid designed so as not to become the winning bid. (The bidder does not want to win this auction, but he or she wants to make sure to be invited to the next auction).

In Australia, a dummy bid (shill, schill) is a criminal offence, but a vendor bid or a co-owner bid below the reserve price is permitted, if clearly declared as such by the auctioneer. These are all official legal terms in Australia, but may have other meanings elsewhere. A co-owner is one of two or several owners (who disagree among themselves).

In Sweden and many other countries there are no legal restrictions, but it will severely hurt the reputation of an auction house that knowingly permits any other bids except genuine bids. If the reserve is not reached this should be clearly declared.

In South Africa auctioneers can use their staff or any bidder to raise the price as long as its disclosed before the auction sale. The Auction Alliance controversy focused on vendor bidding and it was proven to be legal and acceptable in terms of the South African consumer laws.

Suggested Opening Bid (SOB)

There will usually be an estimate of what price the lot will fetch. In an ascending open auction it is considered important to get at least a 50-percent increase in the bids from start to finish. To accomplish this, the auctioneer must start the auction by announcing a suggested opening bid (SOB) that is low enough to be immediately accepted by one of the bidders. Once there is an opening bid, there will quickly be several other, higher bids submitted. Experienced auctioneers will often select an SOB that is about 45 percent of the (lowest) estimate. Thus there is a certain margin of safety to ensure that there will indeed be a lively auction with many bids submitted. Several observations indicate that the lower the SOB, the higher the final winning bid. This is due to the increase in the number of bidders attracted by the low SOB.

A chi-squared distribution shows many low bids but few high bids. Bids "show up together"; without several low bids there will not be any high bids.

Another approach to choosing an SOB: The auctioneer may achieve good success by asking the expected final sales price for the item, as this method suggests to the potential buyers the item's particular value. For instance, say an auctioneer is about to sell a $1,000 car at a sale. Instead of asking $100, hoping to entice wide interest (for who wouldn't want a $1,000 car for $100?), the auctioneer may suggest an opening bid of $1,000; although the first bidder may begin bidding at a mere $100, the final bid may more likely approach $1,000.

Terminology

Duo Yun Xuan auction house in Malacca, Malaysia

- Appraisal – an estimate of an item's worth, usually performed by an expert in that particular field.

- Auction block - a raised platform on which the auctioneer shows the items to be auctioned; can also be slang for the auction itself.

- Auction chant - a rhythmic repetition of numbers and "filler words" spoken by an auctioneer in the process of conducting an auction.

- Auction fever - an emotional state elicited in the course of one or more auctions that causes a bidder to deviate from an initially chosen bidding strategy.

- Auction house - the company operating the auction (i.e., establishing the date and time of the auction, the auction rules, determining which item(s) are to be included in the auction, registering bidders, taking payments, and delivering the goods to the winning bidders).

- Auctioneer - the person conducting the actual auction. They announce the rules of the auction and the item(s) being auctioned, call and acknowledging bids made, and announce the winner. They generally will call the auction using auction chant.

    o The auctioneer may operate his/her own auction house (and thus perform the duties of both auctioneer and auction house), and/or work for another house.

    o Auctioneers are frequently regulated by governmental entities, and in those jurisdictions must meet certain criteria to be licensed (be of a certain age, have no disqualifying criminal record, attend auction school, pass an examination, and pay a licensing fee).

    o Auctioneers may or may not (depending on the laws of the jurisdiction and/or the policies of the auction house) bid for their own account, or if they do must disclose this to bidders at the auction; similar rules may apply for employees of the auctioneer or the auction house.

- Bidding - the act of participating in an auction by offering to purchase an item for sale.

- Buyer's premium – a fee paid by the buyer to the auction house; it is typically calculated as a percentage of the winning bid and added on it. Depending on the jurisdiction the buyer's premium, in addition to the sales price, may be subject to VAT or sales tax.

- Buyout price – A price that, if accepted by a bidder, immediately ends the auction and awards the item to him/her (an example is eBay's BuyItNow feature).

- "Choice" - a form of bidding whereby a number of identical or similar items are bid at a single price *for each item*.

    o "Choice" differs from "lot" in that the winning bidder must take at least one of the items, and can take more than one (up to and including all of them) but is not required to do so.

    o If the bidder takes more than one item, the price paid is "times the money."

    o Items not selected by the winning bidder may then be reauctioned to other bidders.

        ▪ Example: An auction has five bath fragrance gift baskets where bidding is "choice", and the hammer price is USD $5. The winner must choose at least one basket, but can choose two, three, four, or all five baskets. If the winner chooses to take three baskets, s/

he must pay $15 (three baskets @ $5 each). The other two baskets may then be reauctioned.

- Clearance rate – The percentage of items that sell over the course of the auction.

- Commission – a fee paid by a consignor/seller to the auction house; it is typically calculated as a percentage of the winning bid and deducted from the gross proceeds due to the consignor/seller.

- Consignee and consignor - as pertaining to auctions, the consignor (also called the seller, and in some contexts the vendor) is the person owning the item to be auctioned or the owner's representative, while the consignee is the auction house. The consignor maintains title until such time that an item is purchased by a bidder and the bidder pays the auction house.

- Dummy bid (a/k/a "ghost bid") - a false bid, made by someone in collusion with the seller or auctioneer, designed to create a sense of increased interest in the item (and, thus, increased bids).

- Dynamic closing - a mechanism used to prevent auction sniping, by which the closing time is extended for a small period to allow other bidders to increase their bids.

- eBidding – electronic bidding, whereby a person may make a bid without being physically present at an auction (or where the entire auction is taking place on the Internet).

- Earnest money deposit (a/k/a "caution money deposit" or "registration deposit") – a payment that must be made by prospective bidders ahead of time in order to participate in an auction.

    o   The purpose of this deposit is to deter non-serious bidders from attending the auction; by requiring the deposit, only bidders with a genuine interest in the item(s) being sold will participate.

    o   This type of deposit is most often used in auctions involving high-value goods (such as real estate).

    o   The winning bidder has his/her earnest money applied toward the final selling price; the non-winners have theirs refunded to them.

- Escrow – an arrangement in which the winning bidder pays the amount of his/her bid to a third party, who in turn releases the funds to the seller under agreed-upon terms.

- Hammer price – the nominal price at which a lot is sold; the winner is responsible for paying any additional fees and taxes on top of this amount.

- Increment – a minimum amount by which a new bid must exceed the previous bid. An auctioneer may decrease the increment when it appears that bidding on an item may stop, so as to get a higher hammer price. Alternatively, a participant may offer a bid at a smaller increment, which the auctioneer has the discretion to accept or reject.

- Lot – either a single item being sold, or a group of items (which may or may not be similar or identical, such as a "job lot" of manufactured goods) that are bid on as one unit.

    o If the lot is for a group of items, the price paid is for the entire lot and the winning bidder must take all the items sold.

    o Variants on a group lot bid include "choice" and "times the money."

        ▪ Example: An auction has five bath fragrance gift baskets where bidding is "lot", and the hammer price is USD $5. The winner must pay $5 (as the price is for the whole lot) and must take all five baskets.

- Minimum bid – The smallest opening bid that will be accepted.

    o A minimum bid can be as little as USD$0.01 (one cent) depending on the auction.

    o If no one bids at the initial minimum bid, the auctioneer may lower the minimum bid so as to create interest in the item.

    o The minimum bid differs from a reserve price, in that the auctioneer sets the minimum bid, while the seller sets the reserve price (if desired).

- "New money" - a new bidder, joining bidding for an item after others have bid against each other.

- No reserve auction (a/k/a "absolute auction") – an auction in which there is no minimum acceptable price; so long as the winning bid is at least the minimum bid, the seller must honor the sale.

- Outbid (also spelled "out-bid" or "out bid") – to bid higher than another bidder.

- Opening bid – the first bid placed on a particular lot. The opening bid must be at least the minimum bid, but may be higher (e.g., a bidder may shout out a considerably larger bid than minimum, to discourage other bidders from bidding).

- Proxy bid (a/k/a "absentee bid") – a bid placed by an authorized representative of a bidder who is not physically present at the auction.

o   Proxy bids are common in auctions of high-end items, such as art sales (where the proxy represents either a private bidder who does not want to be disclosed to the public, or a museum bidding on a particular item for its collection).

o   If the proxy is outbid on an item during the auction, the proxy (depending on the instructions of the bidder) may either increase the bid (up to a set amount established by the bidder) or be required to drop out of the bidding for that item.

o   A proxy may also be limited by the bidder in the total amount to spend on items in a multi-item auction.

- Relisting - re-selling an item that has already been sold at auction, but where the buyer did not take possession of the item (for example, in a real estate auction, the buyer did not provide payment by the closing date).

- Reserve price – A minimum acceptable price established by the seller prior to the auction, which may or may not be disclosed to the bidders.

o   If the winning bid is below the reserve price, the seller has the right to reject the bid and withdraw the item(s) being auctioned.

o   The reserve price differs from a minimum bid, in that the seller sets the reserve price (if desired), while the auctioneer sets the minimum bid.

- Sealed bid - a submitted bid whose value is unknown to competitors.

- Sniping – the act of placing a bid just before the end of a timed auction, thus giving other bidders no time to enter new bids.

- "Times the money" - a form of bidding similar to "choice", whereby the bid price is *per item*, but where the winning bidder must take *all* of the items offered for sale.

o   The price paid in a "times the money" bid is the bid price multiplied by the number of items, plus buyer's premium and any applicable taxes.

o   "Times the money" differs from "lot" in that the price is *per item*, not one price for all of the items as a group.

- Example: An auction has five bath fragrance gift baskets where bidding is "times the money", and the hammer price is USD $5. The winner must pay $25 (five baskets @ $5 each) and must take all five baskets.

- Vendor bid - a bid by the person selling the item. The bid is sometimes a dummy bid but not always.

JEL Classification

The Journal of Economic Literature (JEL) classification code for auctions is D44.

Selective Breeding

Selective breeding (also called artificial selection) is the process by which humans use animal breeding and plant breeding to selectively develop particular phenotypic traits (characteristics) by choosing which typically animal or plant males and females will sexually reproduce and have offspring together. Domesticated animals are known as breeds, normally bred by a professional breeder, while domesticated plants are known as varieties, cultigens, or cultivars. Two purebred animals of different breeds produce a crossbreed, and crossbred plants are called hybrids. Flowers, vegetables and fruit-trees may be bred by amateurs and commercial or non-commercial professionals: major crops are usually the provenance of the professionals.

A Belgian Blue cow. The defect in the breed's myostatin gene is maintained through linebreeding and is responsible for its accelerated lean muscle growth.

This Chihuahua mix and Great Dane shows the wide range of dog breed sizes created using selective breeding.

There are two approaches or types of artificial selection, or selective breeding. First is the traditional "breeder's approach" in which the breeder or experimenter applies "a known amount of selection to a single phenotypic trait" by examining the chosen trait and choosing to breed only those that exhibit higher or "extreme values" of that trait. The second is called "controlled natural selection," which is essentially natural selection in a controlled environment. In this, the breeder does not choose which individuals being tested "survive or reproduce," as he or she could in the traditional approach. There are also "selection experiments," which is a third approach and these are conducted in order to determine the "strength of natural selection in the wild." However, this is more often an observational approach as opposed to an experimental approach.

Selective breeding transformed teosinte's few fruitcases (left) into modern maize's rows of exposed kernels (right).

In animal breeding, techniques such as inbreeding, linebreeding, and outcrossing are utilized. In plant breeding, similar methods are used. Charles Darwin discussed how selective breeding had been successful in producing change over time in his book, *On the Origin of Species*. The first chapter of the book discusses selective breeding and domestication of such animals as pigeons, cats, cattle, and dogs. Selective breeding was used by Darwin as a springboard to introduce the theory of natural selection, and to support it.

The deliberate exploitation of selective breeding to produce desired results has become very common in agriculture and experimental biology.

Selective breeding can be unintentional, e.g., resulting from the process of human cultivation; and it may also produce unintended – desirable or undesirable – results. For example, in some grains, an increase in seed size may have resulted from certain ploughing practices rather than from the intentional selection of larger seeds. Most likely, there has been an interdependence between natural and artificial factors that have resulted in plant domestication.

History

Selective breeding of both plants and animals has been practiced since early prehistory; key species such as wheat, rice, and dogs have been significantly different from their wild ancestors for millennia, and maize, which required especially large changes from teosinte, its wild form, was selectively bred in Mesoamerica. Selective breeding was practiced by the Romans. Treatises as much as 2,000 years old give advice on selecting animals for different purposes, and these ancient works cite still older authorities, such as Mago the Carthaginian. The notion of selective breeding was later expressed by the

Persian Muslim polymath Abu Rayhan Biruni in the 11th century. He noted the idea in his book titled *India*, and gave various examples.

The agriculturist selects his corn, letting grow as much as he requires, and tearing out the remainder. The forester leaves those branches which he perceives to be excellent, whilst he cuts away all others. The bees kill those of their kind who only eat, but do not work in their beehive.

Selective breeding was established as a scientific practice by Robert Bakewell during the British Agricultural Revolution in the 18th century. Arguably, his most important breeding program was with sheep. Using native stock, he was able to quickly select for large, yet fine-boned sheep, with long, lustrous wool. The Lincoln Longwool was improved by Bakewell, and in turn the Lincoln was used to develop the subsequent breed, named the New (or Dishley) Leicester. It was hornless and had a square, meaty body with straight top lines.

These sheep were exported widely, including to Australia and North America, and have contributed to numerous modern breeds, despite the fact that they fell quickly out of favor as market preferences in meat and textiles changed. Bloodlines of these original New Leicesters survive today as the English Leicester (or Leicester Longwool), which is primarily kept for wool production.

Bakewell was also the first to breed cattle to be used primarily for beef. Previously, cattle were first and foremost kept for pulling ploughs as oxen, but he crossed long-horned heifers and a Westmoreland bull to eventually create the Dishley Longhorn. As more and more farmers followed his lead, farm animals increased dramatically in size and quality. In 1700, the average weight of a bull sold for slaughter was 370 pounds (168 kg). By 1786, that weight had more than doubled to 840 pounds (381 kg). However, after his death, the Dishley Longhorn was replaced with short-horn versions.

He also bred the Improved Black Cart horse, which later became the Shire horse.

Charles Darwin coined the term 'selective breeding'; he was interested in the process as an illustration of his proposed wider process of natural selection. Darwin noted that many domesticated animals and plants had special properties that were developed by intentional animal and plant breeding from individuals that showed desirable characteristics, and discouraging the breeding of individuals with less desirable characteristics.

Darwin used the term "artificial selection" twice in the 1859 first edition of his work *On the Origin of Species*, in Chapter IV: Natural Selection, and in Chapter VI: Difficulties on Theory –

Slow though the process of selection may be, if feeble man can do much by his powers of artificial selection, I can see no limit to the amount of change, to the beauty and infinite complexity of the co-adaptations between all organic beings, one with another and with their physical conditions of life, which may be effected in the long course of time by nature's power of selection.

We are profoundly ignorant of the causes producing slight and unimportant variations; and we are immediately made conscious of this by reflecting on the differences in the breeds of our domesticated animals in different countries,—more especially in the less civilized countries where there has been but little artificial selection.

Animal Breeding

Animals with homogeneous appearance, behavior, and other characteristics are known as particular breeds, and they are bred through culling animals with particular traits and selecting for further breeding those with other traits. Purebred animals have a single, recognizable breed, and purebreds with recorded lineage are called pedigreed. Crossbreeds are a mix of two purebreds, whereas mixed breeds are a mix of several breeds, often unknown. Animal breeding begins with breeding stock, a group of animals used for the purpose of planned breeding. When individuals are looking to breed animals, they look for certain valuable traits in purebred stock for a certain purpose, or may intend to use some type of crossbreeding to produce a new type of stock with different, and, it is presumed, superior abilities in a given area of endeavor. For example, to breed chickens, a typical breeder intends to receive eggs, meat, and new, young birds for further reproduction. Thus, the breeder has to study different breeds and types of chickens and analyze what can be expected from a certain set of characteristics before he or she starts breeding them. Therefore, when purchasing initial breeding stock, the breeder seeks a group of birds that will most closely fit the purpose intended.

Three generations of "Westies" in a village in Fife, Scotland

Purebred breeding aims to establish and maintain stable traits, that animals will pass to the next generation. By "breeding the best to the best," employing a certain degree of inbreeding, considerable culling, and selection for "superior" qualities, one could develop a bloodline superior in certain respects to the original base stock. Such animals can be recorded with a breed registry, the organization that maintains pedigrees and/or stud books. However, single-trait breeding, breeding for only one trait over all others, can be problematic. In one case mentioned by animal behaviorist Temple Grandin, roosters bred for fast growth or heavy muscles did not know how to perform typical rooster courtship dances, which alienated the roosters from hens and led the roosters to kill the hens after reproducing with them.

The observable phenomenon of hybrid vigor stands in contrast to the notion of breed purity. However, on the other hand, indiscriminate breeding of crossbred or hybrid animals may also result in degradation of quality. Studies in evolutionary physiology, behavioral genetics, and other areas of organismal biology have also made use of deliberate selective breeding, though longer generation times and greater difficulty in breeding can make such projects challenging in vertebrates.

Plant Breeding

Plant breeding has been used for thousands of years, and began with the domestication of wild plants into uniform and predictable agricultural cultigens. High-yielding varieties have been particularly important in agriculture.

Researchers at the USDA have selectively bred carrots with a variety of colors.

Selective plant breeding is also used in research to produce transgenic animals that breed "true" (i.e., are homozygous) for artificially inserted or deleted genes.

Selective Breeding in Aquaculture

Selective breeding in aquaculture holds high potential for the genetic improvement of fish and shellfish. Unlike terrestrial livestock, the potential benefits of selective breeding in aquaculture were not realized until recently. This is because high mortality led to the selection of only a few broodstock, causing inbreeding depression, which then forced the use of wild broodstock. This was evident in selective breeding programs for growth rate, which resulted in slow growth and high mortality.

Control of the reproduction cycle was one of the main reasons as it is a requisite for selective breeding programmes. Artificial reproduction was not achieved because of the difficulties in hatching or feeding some farmed species such as eel and yellowtail farming. A suspected reason associated with the late realisation of success in selective breeding programs in aquaculture was the education of the concerned people – researchers, advisory personnel and fish farmers. The education of fish biologists paid less attention to quantitative genetics and breeding plans.

Another was the failure of documentation of the genetic gains in successive generations. This in turn led to failure in quantifying economic benefits that successful selective breeding programs produce. Documentation of the genetic changes was considered important as they help in fine tuning further selection schemes.

Quality Traits in Aquaculture

Aquaculture species are reared for particular traits such as growth rate, survival rate, meat quality, resistance to diseases, age at sexual maturation, fecundity, shell traits like shell size, shell colour, etc.

- Growth rate – growth rate is normally measured as either body weight or body length. This trait is of great economic importance for all aquaculture species as faster growth rate speeds up the turnover of production. Improved growth rates show that farmed animals utilize their feed more efficiently through a correlated response.

- Survival rate – survival rate may take into account the degrees of resistance to diseases. This may also see the stress response as fish under stress are highly vulnerable to diseases. The stress fish experience could be of biological, chemical or environmental influence.

- Meat quality – the quality of fish is of great economic importance in the market. Fish quality usually takes into account size, meatiness, and percentage of fat, colour of flesh, taste, shape of the body, ideal oil and omega-3 content.

- Age at sexual maturation – The age of maturity in aquaculture species is another very important attribute for farmers as during early maturation the species divert all their energy to gonad production affecting growth and meat production and are more susceptible to health problems (Gjerde 1986).

- Fecundity – As the fecundity in fish and shellfish is usually high it is not considered as a major trait for improvement. However, selective breeding practices may consider the size of the egg and correlate it with survival and early growth rate.

Finfish Response to Selection

Salmonids

Gjedrem (1979) showed that selection of Atlantic salmon (*Salmo salar*) led to an increase in body weight by 30% per generation. A comparative study on the performance of select

Atlantic salmon with wild fish was conducted by AKVAFORSK Genetics Centre in Norway. The traits, for which the selection was done included growth rate, feed consumption, protein retention, energy retention, and feed conversion efficiency. Selected fish had a twice better growth rate, a 40% higher feed intake, and an increased protein and energy retention. This led to an overall 20% better Fed Conversion Efficiency as compared to the wild stock. Atlantic salmon have also been selected for resistance to bacterial and viral diseases. Selection was done to check resistance to Infectious Pancreatic Necrosis Virus (IPNV). The results showed 66.6% mortality for low-resistant species whereas the high-resistant species showed 29.3% mortality compared to wild species.

Rainbow trout (*S. gairdneri*) was reported to show large improvements in growth rate after 7–10 generations of selection. Kincaid et al. (1977) showed that growth gains by 30% could be achieved by selectively breeding rainbow trout for three generations. A 7% increase in growth was recorded per generation for rainbow trout by Kause et al. (2005).

In Japan, high resistance to IPNV in rainbow trout has been achieved by selectively breeding the stock. Resistant strains were found to have an average mortality of 4.3% whereas 96.1% mortality was observed in a highly sensitive strain.

Coho salmon (*Oncorhynchus kisutch*) increase in weight was found to be more than 60% after four generations of selective breeding. In Chile, Neira et al. (2006) conducted experiments on early spawning dates in coho salmon. After selectively breeding the fish for four generations, spawning dates were 13–15 days earlier.

Cyprinids

Selective breeding programs for the Common carp (*Cyprinus carpio*) include improvement in growth, shape and resistance to disease. Experiments carried out in the USSR used crossings of broodstocks to increase genetic diversity and then selected the species for traits like growth rate, exterior traits and viability, and/or adaptation to environmental conditions like variations in temperature. Kirpichnikov *et al.* (1974) and Babouchkine (1987) selected carp for fast growth and tolerance to cold, the Ropsha carp. The results showed a 30–40% to 77.4% improvement of cold tolerance but did not provide any data for growth rate. An increase in growth rate was observed in the second generation in Vietnam. Moav and Wohlfarth (1976) showed positive results when selecting for slower growth for three generations compared to selecting for faster growth. Schaperclaus (1962) showed resistance to the dropsy disease wherein selected lines suffered low mortality (11.5%) compared to unselected (57%).

Channel Catfish

Growth was seen to increase by 12–20% in selectively bred *Iictalurus punctatus*. More recently, the response of the Channel Catfish to selection for improved growth rate was found to be approximately 80%, i.e., an average of 13% per generation.

Shellfish Response to Selection

Oysters

Selection for live weight of Pacific oysters showed improvements ranging from 0.4% to 25.6% compared to the wild stock. Sydney-rock oysters (*Saccostrea commercialis*) showed a 4% increase after one generation and a 15% increase after two generations. Chilean oysters (*Ostrea chilensis*), selected for improvement in live weight and shell length showed a 10–13% gain in one generation. Bonamia ostrea is a protistan parasite that causes catastrophic losses (nearly 98%) in European flat oyster *Ostrea edulis* L. This protistan parasite is endemic to three oyster-regions in Europe. Selective breeding programs show that *O. edulis* susceptibility to the infection differs across oyster strains in Europe. A study carried out by Culloty et al. showed that 'Rossmore' oysters in Cork harbour, Ireland had better resistance compared to other Irish strains. A selective breeding program at Cork harbour uses broodstock from 3– to 4-year-old survivors and is further controlled until a viable percentage reaches market size. Over the years 'Rossmore' oysters have shown to develop lower prevalence to *B. ostreae* infection and percentage mortality. Ragone Calvo et al. (2003) selectively bred the eastern oyster, *Crassostrea virginica*, for resistance against co-occurring parasites *Haplosporidium nelson* (MSX) and *Perkinsus marinus* (Dermo). They achieved dual resistance to the disease in four generations of selective breeding. The oysters showed higher growth and survival rates and low susceptibility to the infections. At the end of the experiment, artificially selected *C. virginica* showed a 34–48% higher survival rate.

Penaeid Shrimps

Selection for growth in Penaeid shrimps yielded successful results. A selective breeding program for *Litopenaeus stylirostris* saw an 18% increase in growth after the fourth generation and 21% growth after the fifth generation. *Marsupenaeus japonicas* showed a 10.7% increase in growth after the first generation. Argue et al. (2002) conducted a selective breeding program on the Pacific White Shrimp, *Litopenaeus vannamei* at The Oceanic Institute, Waimanalo, USA from 1995 to 1998. They reported significant responses to selection compared to the unselected control shrimps. After one generation, a 21% increase was observed in growth and 18.4% increase in survival to TSV. The Taura Syndrome Virus (TSV) causes mortalities of 70% or more in shrimps. C.I. Oceanos S.A. in Colombia selected the survivors of the disease from infected ponds and used them as parents for the next generation. They achieved satisfying results in two or three generations wherein survival rates approached levels before the outbreak of the disease. The resulting heavy losses (up to 90%) caused by Infectious hypodermal and haematopoietic necrosis virus (IHHNV) caused a number of shrimp farming industries started to selectively breed shrimps resistant to this disease. Successful outcomes led to development of Super Shrimp, a selected line of *L. stylirostris* that is resistant to IHHNV infection. Tang et al. (2000) confirmed this by showing no mortalities in IHHNV- challenged Super Shrimp post larvae and juveniles.

Aquatic Species Versus Terrestrial Livestock

Selective breeding programs for aquatic species provide better outcomes compared to terrestrial livestock. This higher response to selection of aquatic farmed species can be attributed to the following:

- High fecundity in both sexes fish and shellfish enabling higher selection intensity.

- Large phenotypic and genetic variation in the selected traits.

Selective breeding in aquaculture provide remarkable economic benefits to the industry, the primary one being that it reduces production costs due to faster turnover rates. This is because of faster growth rates, decreased maintenance rates, increased energy and protein retention, and better feed efficiency. Applying such genetic improvement program to aquaculture species will increase productivity to meet the increasing demands of growing populations.

Advantages and Disadvantages

Selective breeding is a direct way to determine if a specific trait can "evolve in response to selection." A single-generation method of breeding is not as accurate or direct. The process is also more practical and easier to understand than sibling analysis. The former tests "differences between line means" while the latter is dependent upon "variance and covariance components." Essentially, selective breeding is better for traits such as physiology and behavior that are hard to measure because it requires fewer individuals to test than single-generation testing.

However, there are disadvantages to this process. Because a single experiment done in selective breeding cannot be used to assess an entire group of "genetic variances and covariances," individual experiments must be done for every individual trait. Also, because of the necessity of selective breeding experiments to require maintaining the organisms tested in a lab or greenhouse, it is impractical to use this breeding method on many organisms. Controlled mating instances are difficult to carry out in this case and this is a necessary component of selective breeding.

Herding

A man herding goats in Tunisia

Herding is the act of bringing individual animals together into a group (herd), maintaining the group, and moving the group from place to place—or any combination of those. Herding can refer either to the process of animals forming herds in the wild, or to human intervention forming herds for some purpose. While the layperson uses the term "herding" to describe this human intervention, most individuals involved in the process term it mustering, "working stock", or droving.

Some animals instinctively gather together as a herd. A group of animals fleeing a predator will demonstrate herd behavior for protection; while some predators, such as wolves and dogs have instinctive herding abilities derived from primitive hunting instincts. Instincts in herding dogs and trainability can be measured at noncompetitive herding tests. Dogs exhibiting basic herding instincts can be trained to aid in herding and to compete in herding and stock dog trials. Sperm whales have also been observed teaming up to herd prey in a coordinated feeding behavior.

Herding is used in agriculture to manage domesticated animals. Herding can be performed by people or trained animals such as herding dogs that control the movement of livestock under the direction of a person. The people whose occupation it is to *herd* or control animals often have *herd* added to the name of the animal they are *herding* to describe their occupation (shepherd, goatherd, cowherd). A competitive sport has developed in some countries where the combined skill of man and herding dog is tested and judged in a *Trial* such as a Sheepdog trial. Animals such as sheep, camel, yak, and goats are mostly reared. They provide milk, meat and other products to the herders and their families.

References

- Hartley, Marie; Ingilby, Joan (1968). Life and Tradition in the Yorkshire Dales. London: J. M. Dent & Sons Ltd. ISBN 0-498-07668-7.

- Michael Pearson; Jane Lennon. Pastoral Australia: Fortunes, Failures and Hard Yakka : a Historical Overview 1788-1967. CSIRO publishing. p. 104. ISBN 9780643096998.

- Mead, edited by G.C. (2004). Poultry meat processing and quality. Cambridge: Woodhead Pub. p. 71. ISBN 978-1-85573-903-1. Retrieved 6 November 2015.

- Milgrom, Paul (2004), Putting Auction Theory to Work, Cambridge, United Kingdom: Cambridge University Press, ISBN 0-521-55184-6

- Shubik, Martin (March 2004), The Theory of Money and Financial Institutions: Volume 1, Cambridge , Mass., USA: MIT Press, pp. 213–219, ISBN 0-262-69311-9

- Buffum, Burt C. (2008). Arid Agriculture; A Hand-Book for the Western Farmer and Stockman. Read Books. p. 232. ISBN 978-1-4086-6710-1.

- Grandin, Temple; Johnson, Catherine (2005). Animals in Translation. New York, New York: Scribner. pp. 69–71. ISBN 0-7432-4769-8.

- Jain, H. K.; Kharkwal, M. C. (2004). Plant breeding - Mendelian to molecular approaches. Boston, London, Dordecht: Kluwer Academic Publishers. ISBN 1-4020-1981-5.

- Gjedrem, T & Baranski, M. (2009). Selective breeding in Aquaculture: An Introduction. 1st Edition. Springer. ISBN 978-90-481-2772-6

- Jones, Sam. "Halal, shechita and the politics of animal slaughter". theguardian.com. The Guardian. Retrieved 2 February 2016.

- British History Online, Memorials of London and London Life in the 13th, 14th and 15th Centuries accessed 23 September 2015

- "Humane Slaughter Association Newsletter March 2011" (PDF). Humane Slaughter Association. Retrieved 1 July 2014.

- "Epic droving trips". The Charleville Times. Brisbane, Queensland: National Library of Australia. 8 February 1951. p. 10. Retrieved 27 February 2013.

- "Remarkable droving feat". The Dubbo Liberal and Macquarie Advocate. New South Wales: National Library of Australia. 11 July 1922. p. 4. Retrieved 15 January 2013.

- "A record droving trip". The Sydney Morning Herald. National Library of Australia. 29 November 1906. p. 5. Retrieved 14 January 2013.

6

Allied Fields of Livestock Production

The fields covered in this chapter are poultry, sericulture and beekeeping. Poultry is the keeping of birds for human purposes such as the production of eggs and meat whereas the maintenance of bees for honey is beekeeping. This section explains to the reader the relation between livestock and its allied fields. It is a compilation of the various branches of livestock production that form an integral part of the broader subject matter.

Poultry

Poultry are domesticated birds kept by humans for the eggs they produce, their meat, their feathers, or sometimes as pets. These birds are most typically members of the superorder Galloanserae (fowl), especially the order Galliformes (which includes chickens, quails and turkeys) and the family Anatidae, in order Anseriformes, commonly known as "waterfowl" and including domestic ducks and domestic geese. Poultry also includes other birds that are killed for their meat, such as the young of pigeons (known as squabs) but does not include similar wild birds hunted for sport or food and known as game. The word "poultry" comes from the French/Norman word *poule*, itself derived from the Latin word *pullus*, which means small animal.

Poultry of the World

The domestication of poultry took place several thousand years ago. This may have originally been as a result of people hatching and rearing young birds from eggs collected from the wild, but later involved keeping the birds permanently in captivity. Domesticated chickens may have been used for cockfighting at first and quail kept for their songs, but soon it was realised how useful it was having a captive-bred source of food. Selective breeding for fast growth, egg-laying ability, conformation, plumage and docility took place over the centuries, and modern breeds often look very different from their wild ancestors. Although some birds are still kept in small flocks in extensive systems, most birds available in the market today are reared in intensive commercial enterprises. Poultry is the second most widely eaten type of meat globally and, along with eggs, provides nutritionally beneficial food containing high-quality protein accompanied by a low proportion of fat. All poultry meat should be properly handled and sufficiently cooked in order to reduce the risk of food poisoning.

Definition

The word "poultry" comes from the Middle English "pultrie", from Old French *pouletrie*, from *pouletier*, poultry dealer, from *poulet*, pullet. The word "pullet" itself comes from Middle English *pulet*, from Old French *polet*, both from Latin *pullus*, a young fowl, young animal or chicken. The word "fowl" is of Germanic origin (cf. Old English *Fugol*, German *Vogel*, Danish *Fugl*).

"Poultry" is a term used for any kind of domesticated bird, captive-raised for its utility, and traditionally the word has been used to refer to wildfowl (Galliformes) and waterfowl (Anseriformes). "Poultry" can be defined as domestic fowls, including chickens, turkeys, geese and ducks, raised for the production of meat or eggs and the word is also used for the flesh of these birds used as food. The Encyclopædia Britannica lists the same bird groups but also includes guinea fowl and squabs (young pigeons). In R. D. Crawford's *Poultry breeding and genetics*, squabs are omitted but Japanese quail and common pheasant are added to the list, the latter frequently being bred in captivity and released into the wild. In his 1848 classic book on poultry, *Ornamental and Domestic Poultry: Their History, and Management*, Edmund Dixon included chapters on the peafowl, guinea fowl, mute swan, turkey, various types of geese, the muscovy duck, other ducks and all types of chickens including bantams. In colloquial speech, the term "fowl" is often used near-synonymously with "domesticated chicken" (*Gallus gallus*), or with "poultry" or even just "bird", and many languages do not distinguish between "poultry" and "fowl". Both words are also used for the flesh of these birds. Poultry can be distinguished from "game", defined as wild birds or mammals hunted for food or sport, a word also used to describe the flesh of these when eaten.

Examples

Bird	Wild ancestor	Domestication	Utilization	
Chicken	Red junglefowl	Southeast Asia	meat, feathers, eggs, ornamentation, leather	

Duck	Muscovy duck/ Mallard	various	meat, feathers, eggs	
Emu	Emu	various, 20th century	meat, leather, oil	
Goose	Greylag goose/ Swan goose	various	meat, feathers, eggs	
Indian peafowl	Indian Peafowl	various	meat, feathers, ornamentation, landscaping	
Mute swan	Mute swan	various	feathers, eggs, landscaping	
Ostrich	Ostrich	various, 20th century	meat, eggs, feathers, leather	
Pigeon	Rock dove	Middle East	meat, feathers, ornamentation	
Quail	Japanese quail	Japan	meat, eggs, feathers, pets	

Turkey	Wild turkey	Mexico	meat, feathers	
Guineafowl	Helmeted guineafowl	Africa	meat, pest consumption, and alarm calling	
Common pheasant	Common pheasant	Eurasia	meat	
Golden pheasant	Golden pheasant	Eurasia	meat, mainly ornamental	
Rhea	Rhea	various, 20th century	meat, leather, oil, eggs	

Chickens

Chickens are medium-sized, chunky birds with an upright stance and characterised by fleshy red combs and wattles on their heads. Males, known as cocks, are usually larger, more boldly coloured, and have more exaggerated plumage than females (hens). Chickens are gregarious, omnivorous, ground-dwelling birds that in their natural surroundings search among the leaf litter for seeds, invertebrates, and other small animals. They seldom fly except as a result of perceived danger, preferring to run into the undergrowth if approached. Today's domestic chicken (*Gallus gallus domesticus*) is mainly descended from the wild red junglefowl of Asia, with some additional input from grey junglefowl. Domestication is believed to have taken place between 7,000 and 10,000 years ago, and what are thought to be fossilized chicken bones have been found in northeastern China dated to around 5,400 BC. Archaeologists believe domestication was originally for the purpose of cockfighting, the male bird being a doughty fighter. By 4,000 years ago, chickens seem to have reached the Indus Valley and 250 years later, they arrived in Egypt. They were still used for fighting and were regarded as symbols of fertility. The Romans used them in divination, and the Egyptians made a breakthrough when they learned the difficult technique of artificial incubation. Since then, the keeping of chickens has spread around the world for the production of food with the domestic fowl being a valuable source of both eggs and meat.

Cock with comb and wattles

Since their domestication, a large number of breeds of chickens have been established, but with the exception of the white Leghorn, most commercial birds are of hybrid origin. In about 1800, chickens began to be kept on a larger scale, and modern high-output poultry farms were present in the United Kingdom from around 1920 and became established in the United States soon after the Second World War. By the mid-20th century, the poultry meat-producing industry was of greater importance than the egg-laying industry. Poultry breeding has produced breeds and strains to fulfil different needs; light-framed, egg-laying birds that can produce 300 eggs a year; fast-growing, fleshy birds destined for consumption at a young age, and utility birds which produce both an acceptable number of eggs and a well-fleshed carcase. Male birds are unwanted in the egg-laying industry and can often be identified as soon as they are hatch for subsequent culling. In meat breeds, these birds are sometimes castrated (often chemically) to prevent aggression. The resulting bird, called a capon, has more tender and flavorful meat, as well.

Roman mosaic depicting a cockfight

A bantam is a small variety of domestic chicken, either a miniature version of a member of a standard breed, or a "true bantam" with no larger counterpart. The name derives from the town of Bantam in Java where European sailors bought the local small chickens for their shipboard supplies. Bantams may be a quarter to a third of the size of standard birds and lay similarly small eggs. They are kept by small-holders and hobbyists for egg production, use as broody hens, ornamental purposes, and showing.

Cockfighting

Cockfighting is said to be the world's oldest spectator sport and may have originated in Persia 6,000 years ago. Two mature males (cocks or roosters) are set to fight each other, and will do so with great vigour until one is critically injured or killed. Breeds such as the Aseel were developed in the Indian subcontinent for their aggressive behaviour. The sport formed part of the culture of the ancient Indians, Chinese, Greeks, and Romans, and large sums were won or lost depending on the outcome of an encounter. Cockfighting has been banned in many countries during the last century on the grounds of cruelty to animals.

Ducks

Ducks are medium-sized aquatic birds with broad bills, eyes on the side of the head, fairly long necks, short legs set far back on the body, and webbed feet. Males, known as drakes, are often larger than females (simply known as ducks) and are differently coloured in some breeds. Domestic ducks are omnivores, eating a variety of animal and plant materials such as aquatic insects, molluscs, worms, small amphibians, waterweeds, and grasses. They feed in shallow water by dabbling, with their heads underwater and their tails upended. Most domestic ducks are too heavy to fly, and they are social birds, preferring to live and move around together in groups. They keep their plumage waterproof by preening, a process that spreads the secretions of the preen gland over their feathers.

Pekin ducks

Clay models of ducks found in China dating back to 4000 BC may indicate the domestication of ducks took place there during the Yangshao culture. Even if this is not the case, domestication of the duck took place in the Far East at least 1500 years earlier than in the West. Lucius Columella, writing in the first century BC, advised those who sought to rear ducks to collect wildfowl eggs and put them under a broody hen, because when raised in this way, the ducks "lay aside their wild nature and without hesitation breed when shut up in the bird pen". Despite this, ducks did not appear in agricultural texts in Western Europe until about 810 AD, when they began to be mentioned alongside geese, chickens, and peafowl as being used for rental payments made by tenants to landowners.

It is widely agreed that the mallard (*Anas platyrhynchos*) is the ancestor of all breeds of domestic duck (with the exception of the Muscovy duck (*Cairina moschata*), which is not closely related to other ducks). Ducks are farmed mainly for their meat, eggs, and down. As is the case with chickens, various breeds have been developed, selected for egg-laying ability, fast growth, and a well-covered carcase. The most common commercial breed in the United Kingdom and the United States is the Pekin duck, which can lay 200 eggs a year and can reach a weight of 3.5 kg (7.7 lb) in 44 days. In the Western world, ducks are not as popular as chickens, because the latter produce larger quantities of white, lean meat and are easier to keep intensively, making the price of chicken meat lower than that of duck meat. While popular in *haute cuisine*, duck appears less frequently in the mass-market food industry. However, things are different in the East. Ducks are more popular there than chickens and are mostly still herded in the traditional way and selected for their ability to find sufficient food in harvested rice fields and other wet environments.

Geese

The greylag goose (*Anser anser*) was domesticated by the Egyptians at least 3000 years ago, and a different wild species, the swan goose (*Anser cygnoides*), domesticated in Siberia about a thousand years later, is known as a Chinese goose. The two hybridise with each other and the large knob at the base of the beak, a noticeable feature of the Chinese goose, is present to a varying extent in these hybrids. The hybrids are fertile and have resulted in several of the modern breeds. Despite their early domestication, geese have never gained the commercial importance of chickens and ducks.

An Emden goose, a descendent of the wild greylag goose

Domestic geese are much larger than their wild counterparts and tend to have thick necks, an upright posture, and large bodies with broad rear ends. The greylag-derived birds are large and fleshy and used for meat, while the Chinese geese have smaller frames and are mainly used for egg production. The fine down of both is valued for use in pillows and padded garments. They forage on grass and weeds, supplementing this with small invertebrates, and one of the attractions of rearing geese is their ability to grow and thrive on a grass-based system. They are very gregarious and have good memories and can be allowed to roam widely in the knowledge that they will return home by dusk. The Chinese goose is more aggressive and noisy than other geese and can be used as a guard animal to warn of intruders. The flesh of meat geese is dark-coloured and high in protein, but they deposit fat subcutaneously, although this fat contains mostly monounsaturated fatty acids. The birds are killed either around 10 or about 24 weeks. Between these ages, problems with dressing the carcase occur because of the presence of developing pin feathers.

In some countries, geese and ducks are force-fed to produce livers with an exceptionally high fat content for the production of *foie gras*. Over 75% of world production of this product occurs in France, with lesser industries in Hungary and Bulgaria and a growing production in China. *Foie gras* is considered a luxury in many parts of the world, but the process of feeding the birds in this way is banned in many countries on animal welfare grounds.

Turkeys

Turkeys are large birds, their nearest relatives being the pheasant and the guineafowl. Males are larger than females and have spreading, fan-shaped tails and distinctive, fleshy wattles, called a snood, that hang from the top of the beak and are used in courtship display. Wild turkeys can fly, but seldom do so, preferring to run with a long, stratling gait. They roost in trees and forage on the ground, feeding on seeds, nuts, berries, grass, foliage, invertebrates, lizards, and small snakes.

Male domesticated turkey sexually displaying by showing the snood hanging over the beak, the caruncles hanging from the throat, and the 'beard' of small, black, stiff feathers on the chest

The modern domesticated turkey is descended from one of six subspecies of wild turkey (*Meleagris gallopavo*) found in the present Mexican states of Jalisco, Guerrero and Veracruz. Pre-Aztec tribes in south-central Mexico first domesticated the bird around

800 BC, and Pueblo Indians inhabiting the Colorado Plateau in the United States did likewise around 200 BC. They used the feathers for robes, blankets, and ceremonial purposes. More than 1,000 years later, they became an important food source. The first Europeans to encounter the bird misidentified it as a guineafowl, a bird known as a "turkey fowl" at that time because it had been introduced into Europe via Turkey.

Commercial turkeys are usually reared indoors under controlled conditions. These are often large buildings, purpose-built to provide ventilation and low light intensities (this reduces the birds' activity and thereby increases the rate of weight gain). The lights can be switched on for 24-hrs/day, or a range of step-wise light regimens to encourage the birds to feed often and therefore grow rapidly. Females achieve slaughter weight at about 15 weeks of age and males at about 19. Mature commercial birds may be twice as heavy as their wild counterparts. Many different breeds have been developed, but the majority of commercial birds are white, as this improves the appearance of the dressed carcass, the pin feathers being less visible. Turkeys were at one time mainly consumed on special occasions such as Christmas (10 million birds in the United Kingdom) or Thanksgiving (60 million birds in the United States). However, they are increasingly becoming part of the everyday diet in many parts of the world.

Quail

The quail is a small to medium-sized, cryptically coloured bird. In its natural environment, it is found in bushy places, in rough grassland, among agricultural crops, and in other places with dense cover. It feeds on seeds, insects, and other small invertebrates. Being a largely ground-dwelling, gregarious bird, domestication of the quail was not difficult, although many of its wild instincts are retained in captivity. It was known to the Egyptians long before the arrival of chickens and was depicted in hieroglyphs from 2575 BC. It migrated across Egypt in vast flocks and the birds could sometimes be picked up off the ground by hand. These were the common quail (*Coturnix coturnix*), but modern domesticated flocks are mostly of Japanese quail (*Coturnix japonica*) which was probably domesticated as early as the 11th century AD in Japan. They were originally kept as songbirds, and they are thought to have been regularly used in song contests.

Japanese quail

In the early 20th century, Japanese breeders began to selectively breed for increased egg production. By 1940, the quail egg industry was flourishing, but the events of World War II led to the complete loss of quail lines bred for their song type, as well as almost all of those bred for egg production. After the war, the few surviving domesticated quail were used to rebuild the industry, and all current commercial and laboratory lines are considered to have originated from this population. Modern birds can lay upward of 300 eggs a year and countries such as Japan, India, China, Italy, Russia, and the United States have established commercial Japanese quail farming industries. Japanese quail are also used in biomedical research in fields such as genetics, embryology, nutrition, physiology, pathology, and toxicity studies. These quail are closely related to the common quail, and many young hybrid birds are released into the wild each year to replenish dwindling wild populations.

Other Poultry

Guinea fowl originated in southern Africa, and the species most often kept as poultry is the helmeted guineafowl (*Numida meleagris*). It is a medium-sized grey or speckled bird with a small naked head with colourful wattles and a knob on top, and was domesticated by the time of the ancient Greeks and Romans. Guinea fowl are hardy, sociable birds that subsist mainly on insects, but also consume grasses and seeds. They will keep a vegetable garden clear of pests and will eat the ticks that carry Lyme disease. They happily roost in trees and give a loud vocal warning of the approach of predators. Their flesh and eggs can be eaten in the same way as chickens, young birds being ready for the table at the age of about four months.

A squab is the name given to the young of domestic pigeons that are destined for the table. Like other domesticated pigeons, birds used for this purpose are descended from the rock pigeon (*Columba livia*). Special utility breeds with desirable characteristics are used. Two eggs are laid and incubated for about 17 days. When they hatch, the squabs are fed by both parents on "pigeon's milk", a thick secretion high in protein produced by the crop. Squabs grow rapidly, but are slow to fledge and are ready to leave the nest at 26 to 30 days weighing about 500 g (18 oz). By this time, the adult pigeons will have laid and be incubating another pair of eggs and a prolific pair should produce two squabs every four weeks during a breeding season lasting several months.

Poultry Farming

Worldwide, more chickens are kept than any other type of poultry, with over 50 billion birds being raised each year as a source of meat and eggs. Traditionally, such birds would have been kept extensively in small flocks, foraging during the day and housed at night. This is still the case in developing countries, where the women often make important contributions to family livelihoods through keeping poultry. However, rising world populations and urbanization have led to the bulk of production being in larger, more intensive specialist units. These are often situated close to where the feed is grown or near to where the meat is needed, and result in cheap, safe food being made available for urban communities. Profitability of production depends very much on the price of feed, which has been rising. High feed costs could limit further development of poultry production.

Free-range ducks in Hainan Province, China

In free-range husbandry, the birds can roam freely outdoors for at least part of the day. Often, this is in large enclosures, but the birds have access to natural conditions and can exhibit their normal behaviours. A more intensive system is yarding, in which the birds have access to a fenced yard and poultry house at a higher stocking rate. Poultry can also be kept in a barn system, with no access to the open air, but with the ability to move around freely inside the building. The most intensive system for egg-laying chickens is battery cages, often set in multiple tiers. In these, several birds share a small cage which restricts their ability to move around and behave in a normal manner. The eggs are laid on the floor of the cage and roll into troughs outside for ease of collection. Battery cages for hens have been illegal in the EU since January 1, 2012.

Chickens raised intensively for their meat are known as "broilers". Breeds have been developed that can grow to an acceptable carcass size (2 kg (4.4 lb)) in six weeks or less. Broilers grow so fast, their legs cannot always support their weight and their hearts and respiratory systems may not be able to supply enough oxygen to their developing muscles. Mortality rates at 1% are much higher than for less-intensively reared laying birds which take 18 weeks to reach similar weights. Processing the birds is done automatically with conveyor-belt efficiency. They are hung by their feet, stunned, killed, bled, scalded, plucked, have their heads and feet removed, eviscerated, washed, chilled, drained, weighed, and packed, all within the course of little over two hours.

Both intensive and free-range farming have animal welfare concerns. In intensive systems, cannibalism, feather pecking and vent pecking can be common, with some farmers using beak trimming as a preventative measure. Diseases can also be common and spread rapidly through the flock. In extensive systems, the birds are exposed to adverse weather conditions and are vulnerable to predators and disease-carrying wild birds. Barn systems have been found to have the worst bird welfare. In Southeast Asia, a lack of disease control in free-range farming has been associated with outbreaks of avian influenza.

Poultry Shows

In many countries, national and regional poultry shows are held where enthusiasts exhibit their birds which are judged on certain phenotypical breed traits as specified by their respective breed standards. The idea of poultry exhibition may have originated after cockfighting was made illegal, as a way of maintaining a competitive element in poultry husbandry. Breed standards were drawn up for egg-laying, meat-type, and purely ornamental birds, aiming for uniformity. Sometimes, poultry shows are part of general livestock shows, and sometimes they are separate events such as the annual "National Championship Show" in the United Kingdom organised by the Poultry Club of Great Britain.

Poultry as Food

Trade

Chicken and duck eggs on sale in Hong Kong

Poultry is the second most widely eaten type of meat in the world, accounting for about 30% of total meat production worldwide compared to pork at 38%. Sixteen billion birds are raised annually for consumption, more than half of these in industrialised, factory-like production units. Global broiler meat production rose to 84.6 million tonnes in 2013. The largest producers were the United States (20%), China (16.6%), Brazil (15.1%) and the European Union (11.3%). There are two distinct models of production; the European Union supply chain model seeks to supply products which can be traced back to the farm of origin. This model faces the increasing costs of implementing additional food safety requirements, welfare issues and environmental regulations. In contrast, the United States model turns the product into a commodity.

World production of duck meat was about 4.2 million tonnes in 2011 with China producing two thirds of the total, some 1.7 billion birds. Other notable duck-producing countries in the Far East include Vietnam, Thailand, Malaysia, Myanmar, Indonesia and South Korea (12% in total). France (3.5%) is the largest producer in the West, followed by other EU nations (3%) and North America (1.7%). China was also by far the largest producer of goose and guinea fowl meat, with a 94% share of the 2.6 million tonne global market.

Global egg production was expected to reach 65.5 million tonnes in 2013, surpassing all previous years. Between 2000 and 2010, egg production was growing globally at around 2% per year, but since then growth has slowed down to nearer 1%.

Cuts of Poultry

Poultry is available fresh or frozen, as whole birds or as joints (cuts), bone-in or deboned, seasoned in various ways, raw or ready cooked. The meatiest parts of a bird are the flight muscles on its chest, called "breast" meat, and the walking muscles on the legs, called the "thigh" and "drumstick". The wings are also eaten (Buffalo wings are a popular example in the United States) and may be split into three segments, the meatier "drumette", the "wingette" (also called the "flat"), and the wing tip (also called the "flapper"). In Japan, the wing is frequently separated, and these parts are referred to as 手羽元 (*teba-moto* "wing base") and 手羽先 (*teba-saki* "wing tip").

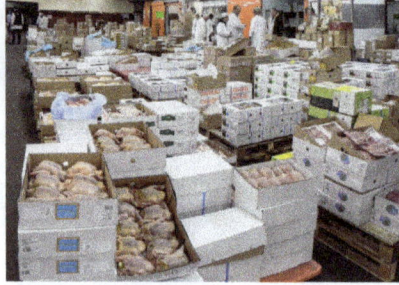

In the poultry pavilion of the Rungis International Market, France

Dark meat, which avian myologists refer to as "red muscle", is used for sustained activity—chiefly walking, in the case of a chicken. The dark colour comes from the protein myoglobin, which plays a key role in oxygen uptake and storage within cells. White muscle, in contrast, is suitable only for short bursts of activity such as, for chickens, flying. Thus, the chicken's leg and thigh meat are dark, while its breast meat (which makes up the primary flight muscles) is white. Other birds with breast muscle more suitable for sustained flight, such as ducks and geese, have red muscle (and therefore dark meat) throughout. Some cuts of meat including poultry expose the microscopic regular structure of intracellular muscle fibrils which can diffract light and produce iridescent colours, an optical phenomenon sometimes called structural colouration.

Health and Disease (Humans)

Poultry meat and eggs provide nutritionally beneficial food containing protein of high quality. This is accompanied by low levels of fat which have a favourable mix of fatty acids. Chicken meat contains about two to three times as much polyunsaturated fat as most types of red meat when measured by weight. However, for boneless, skinless chicken breast, the amount is much lower. A 100-g serving of baked chicken breast contains 4 g of fat and 31 g of protein, compared to 10 g of fat and 27 g of protein for the same portion of broiled, lean skirt steak.

Cuts from plucked chickens

A 2011 study by the Translational Genomics Research Institute showed that 47% of the meat and poultry sold in United States grocery stores was contaminated with *Staphylococcus aureus*, and 52% of the bacteria concerned showed resistance to at least three groups of antibiotics. Thorough cooking of the product would kill these bacteria, but a risk of cross-contamination from improper handling of the raw product is still present. Also, some risk is present for consumers of poultry meat and eggs to bacterial infections such as *Salmonella* and *Campylobacter*. Poultry products may become contaminated by these bacteria during handling, processing, marketing, or storage, resulting in food-borne illness if the product is improperly cooked or handled.

In general, avian influenza is a disease of birds caused by bird-specific influenza A virus that is not normally transferred to people; however, people in contact with live poultry are at the greatest risk of becoming infected with the virus and this is of particular concern in areas such as Southeast Asia, where the disease is endemic in the wild bird population and domestic poultry can become infected. The virus possibly could mutate to become highly virulent and infectious in humans and cause an influenza pandemic.

Bacteria can be grown in the laboratory on nutrient culture media, but viruses need living cells in which to replicate. Many vaccines to infectious diseases can be grown in fertilised chicken eggs. Millions of eggs are used each year to generate the annual flu vaccine requirements, a complex process that takes about six months after the decision is made as to what strains of virus to include in the new vaccine. A problem with using eggs for this purpose is that people with egg allergies are unable to be immunised, but this disadvantage may be overcome as new techniques for cell-based rather than egg-based culture become available. Cell-based culture will also be useful in a pandemic when it may be difficult to acquire a sufficiently large quantity of suitable sterile, fertile eggs.

Sericulture

silkworm and cocoon

Sericulture, or silk farming, is the rearing of silkworms for the production of silk. Although there are several commercial species of silkworms, *Bombyx mori* (the caterpillar of the domesticated silk moth) is the most widely used and intensively studied silkworm. Silk was first produced in China as early as the Neolithic period. Sericulture has become an important cottage industry in countries such as Brazil, China, France, India, Italy, Japan, Korea, and Russia. Today, China and India are the two main producers, with more than 60% of the world's annual production.

History

According to Confucian text, the discovery of silk production dates to about 2700 BC, although archaeological records point to silk cultivation as early as the Yangshao period (5000 – 3000 BC). By about the first half of the 1st century AD it had reached ancient Khotan, by a series of interactions along the Silk Road. By 140 AD the practice had been established in India. In the 6th century the smuggling of silkworm eggs into the Byzantine Empire led to its establishment in the Mediterranean, remaining a monopoly in the Byzantine Empire for centuries (Byzantine silk). In 1147, during the Second Crusade, Roger II of Sicily (1095–1154) attacked Corinth and Thebes, two important centres of Byzantine silk production, capturing the weavers and their equipment and establishing his own silkworks in Palermo and Calabria, eventually spreading the industry to Western Europe.

- Chinese sericulture process

- The silkworms and mulberry leaves are placed on trays.

- Twig frames for the silkworms are prepared.

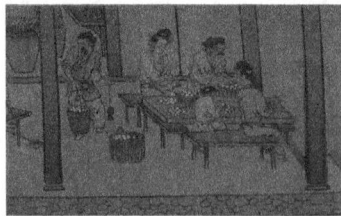

The cocoons are weighed.

The cocoons are soaked and the silk is wound on spools.

The silk is woven using a loom.

Production

Silkworm larvae are fed with mulberry leaves, and, after the fourth moult, climb a twig placed near them and spin their silken cocoons. This process is achieved by the worm through a dense fluid secreted from its structural glands, resulting in the fiber of the cocoon. The silk is a continuous filament comprising fibroin protein, secreted from two salivary glands in the head of each larva, and a gum called sericin, which cements the filaments. The sericin is removed by placing the cocoons in hot water, which frees the silk filaments and readies them for reeling. This is known as the degumming process. The immersion in hot water also kills the silkworm pupae.

Single filaments are combined to form thread, which is drawn under tension through several guides and wound onto reels. The threads may be plied to form yarn. After drying, the raw silk is packed according to quality.

Stages of Production

The stages of production are as follows:

1. The silk moth lays thousands of eggs.

2. The silk moth eggs hatch to form larvae or caterpillars, known as silkworms.

3. The larvae feed on mulberry leaves.

4. Having grown and moulted several times silkworm weaves a net to hold itself

5. It swings its head from side to side in a figure '8' distributing the saliva that will form silk.

6. The silk solidifies when it contacts the air.

7. The silkworm spins approximately one mile of filament and completely encloses itself in a cocoon in about two or three days. The amount of usable quality silk in each cocoon is small. As a result, about 2500 silkworms are required to produce a pound of raw silk

8. The intact cocoons are boiled, killing the silkworm pupae.

9. The silk is obtained by brushing the undamaged cocoon to find the outside end of the filament.

10. The silk filaments are then wound on a reel. One cocoon contains approximately 1,000 yards of silk filament. The silk at this stage is known as raw silk. One thread comprises up to 48 individual silk filaments.

Mahatma Gandhi was critical of silk production based on the Ahimsa philosophy "not to hurt any living thing". He also promoted 'Ahimsa silk', made without boiling the pupae to procure the silk and wild silk made from the cocoons of wild and semi-wild silk moths. In the early 21st century the organisation PETA has campaigned against silk.

third stage of silkworm

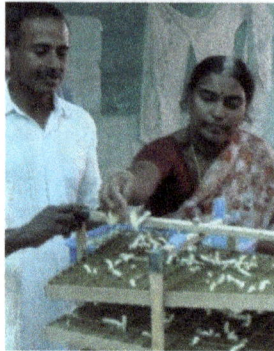

silkworms on to Modern Rotary mountage

Beekeeping

Apiculture (from Latin: *apis* "bee") is the maintenance of honey bee colonies, commonly in hives, by humans. A beekeeper (or apiarist) keeps bees in order to collect their honey and other products that the hive produces (including beeswax, propolis, pollen, and royal jelly), to pollinate crops, or to produce bees for sale to other beekeepers. A location where bees are kept is called an apiary or "bee yard".

Beekeeping, tacuinum sanitatis casanatensis (14th century)

Beekeeping in Serbia

Honey seeker depicted on 8000-year-old cave painting near Valencia, Spain

Depictions of humans collecting honey from wild bees date to 15,000 years ago. Bee-keeping in pottery vessels began about 9,000 years ago in North Africa. Domestication is shown in Egyptian art from around 4,500 years ago. Simple hives and smoke were used and honey was stored in jars, some of which were found in the tombs of pharaohs such as Tutankhamun. It wasn't until the 18th century that European understanding of the colonies and biology of bees allowed the construction of the moveable comb hive so that honey could be harvested without destroying the entire colony.

History of Beekeeping

At some point humans began to attempt to domesticate wild bees in artificial hives made from hollow logs, wooden boxes, pottery vessels, and woven straw baskets or "skeps". Traces of beeswax are found in pot sherds throughout the Middle East beginning about 7000 BCE.

Honeybees were kept in Egypt from antiquity. On the walls of the sun temple of Nyus-erre Ini from the Fifth Dynasty, before 2422 BCE, workers are depicted blowing smoke into hives as they are removing honeycombs. Inscriptions detailing the production of honey are found on the tomb of Pabasa from the Twenty-sixth Dynasty (c. 650 BCE), depicting pouring honey in jars and cylindrical hives. Sealed pots of honey were found in the grave goods of pharaohs such as Tutankhamun.

There was a documented attempt to introduce bees to dry areas of Mesopotamia in the 8th century BCE by Shamash-resh-uşur, the governor of Mari and Suhu. His plans were detailed in a stele of 760 BCE:

Stele showing Shamash-resh-uşur praying to the gods Adad and Ishtar with an inscription in Babylonian cuneiform.

I am Shamash-resh-uşur , the governor of Suhu and the land of Mari. Bees that collect honey, which none of my ancestors had ever seen or brought into the land of Suhu, I brought down from the mountain of the men of Habha, and made them settle in the orchards of the town 'Gabbari-built-it'. They collect honey and wax, and I know how to melt the honey and wax – and the gardeners know too. Whoever comes in the future, may he ask the old men of the town, (who will say) thus: "They are the buildings of Shamash-resh-uşur, the governor of Suhu, who introduced honey bees into the land of Suhu."

— translated text from stele, (Dalley, 2002)

In prehistoric Greece (Crete and Mycenae), there existed a system of high-status apiculture, as can be concluded from the finds of hives, smoking pots, honey extractors and other beekeeping paraphernalia in Knossos. Beekeeping was considered a highly valued industry controlled by beekeeping overseers—owners of gold rings depicting apiculture scenes rather than religious ones as they have been reinterpreted recently, contra Sir Arthur Evans.

Archaeological finds relating to beekeeping have been discovered at Rehov, a Bronze and Iron Age archaeological site in the Jordan Valley, Israel. Thirty intact hives, made of straw and unbaked clay, were discovered by archaeologist Amihai Mazar in the ruins of the city, dating from about 900 BCE. The hives were found in orderly rows, three high, in a manner that could have accommodated around 100 hives, held more than 1 million bees and had a potential annual yield of 500 kilograms of honey and 70 kilograms of beeswax, according to Mazar, and are evidence that an advanced honey industry existed in ancient Israel 3,000 years ago.

The Beekeepers, 1568, by Pieter Bruegel the Elder

In ancient Greece, aspects of the lives of bees and beekeeping are discussed at length by Aristotle. Beekeeping was also documented by the Roman writers Virgil, Gaius Julius Hyginus, Varro, and Columella.

Beekeeping has also been practiced in ancient China since antiquity. In the book "Golden Rules of Business Success" written by Fan Li (or Tao Zhu Gong) during the Spring and Autumn Period there are sections describing the art of beekeeping, stressing the importance of the quality of the wooden box used and how this can affect the quality of the honey.

The ancient Maya domesticated a separate species of stingless bee. The use of stingless bees is referred to as meliponiculture, named after bees of the tribe Meliponini—such as *Melipona quadrifasciata* in Brazil. This variation of bee keeping still occurs around the world today. For instance, in Australia, the stingless bee *Tetragonula carbonaria* is kept for production of their honey.

Origins

There are more than 20,000 species of wild bees. Many species are solitary (e.g., mason bees, leafcutter bees (Megachilidae), carpenter bees and other ground-nesting bees). Many others rear their young in burrows and small colonies (e.g., bumblebees and stingless bees). Some honey bees are wild e.g. the little honeybee (*Apis florea*), giant honeybee (*Apis dorsata*) and rock bee (*Apis laboriosa*). Beekeeping, or apiculture, is concerned with the practical management of the social species of honey bees, which live in large colonies of up to 100,000 individuals. In Europe and America the species universally managed by beekeepers is the Western honey bee (*Apis mellifera*). This species has several sub-species or regional varieties, such as the Italian bee (*Apis mellifera ligustica*), European dark bee (*Apis mellifera mellifera*), and the Carniolan honey bee (*Apis mellifera carnica*). In the tropics, other species of social bees are managed for honey production, including the Asiatic honey bee (*Apis cerana*).

All of the *Apis mellifera* sub-species are capable of inter-breeding and hybridizing. Many bee breeding companies strive to selectively breed and hybridize varieties to produce desirable qualities: disease and parasite resistance, good honey production, swarming behaviour reduction, prolific breeding, and mild disposition. Some of these hybrids are marketed under specific brand names, such as the Buckfast Bee or Midnite Bee. The advantages of the initial F1 hybrids produced by these crosses include: hybrid vigor, increased honey productivity, and greater disease resistance. The disadvantage is that in subsequent generations these advantages may fade away and hybrids tend to be very defensive and aggressive.

Wild honey Harvesting

Collecting honey from wild bee colonies is one of the most ancient human activities and is still practiced by aboriginal societies in parts of Africa, Asia, Australia, and South America. In Africa, honeyguide birds have evolved a mutualist relationship with humans, leading them to hives and participating in the feast. This suggests honey harvesting by humans may be of great antiquity. Some of the earliest evidence of gathering honey from wild colonies is from rock paintings, dating to around Upper Paleolithic (13,000 BCE). Gathering honey from wild bee colonies is usually done by subduing the bees with smoke and breaking open the tree or rocks where the colony is located, often resulting in the physical destruction of the nest.

Wild bees' nest, suspended from a branch

Study of Honey Bees

It was not until the 18th century that European natural philosophers undertook the scientific study of bee colonies and began to understand the complex and hidden world of bee biology. Preeminent among these scientific pioneers were Swammerdam, René Antoine Ferchault de Réaumur, Charles Bonnet, and Francois Huber. Swammerdam and Réaumur were among the first to use a microscope and dissection to understand the internal biology of honey bees. Réaumur was among the first to construct a glass walled observation hive to better observe activities within hives. He observed queens laying eggs in open cells, but still had no idea of how a queen was fertilized; nobody had ever witnessed the mating of a queen and drone and many theories held that queens were "self-fertile," while others believed that a vapor or "miasma" emanating from the drones fertilized queens without direct physical contact. Huber was the first to prove by observation and experiment that queens are physically inseminated by drones outside the confines of hives, usually a great distance away.

Following Réaumur's design, Huber built improved glass-walled observation hives and sectional hives that could be opened like the leaves of a book. This allowed inspecting individual wax combs and greatly improved direct observation of hive activity. Although he went blind before he was twenty, Huber employed a secretary, Francois Burnens, to make daily observations, conduct careful experiments, and keep accurate notes over more than twenty years. Huber confirmed that a hive consists of one queen who is the mother of all the female workers and male drones in the colony. He was also the first to confirm that mating with drones takes place outside of hives and that queens are inseminated by a number of successive matings with male drones, high in the air at a great distance from their hive. Together, he and Burnens dissected bees under the microscope and were among the first to describe the ovaries and spermatheca, or sperm store, of queens as well as the penis of male drones. Huber is universally regarded as "the father of modern bee-science" and his "Nouvelles Observations sur Les Abeilles (or "New Observations on Bees") revealed all the basic scientific truths for the biology and ecology of honeybees.

Invention of the Movable Comb Hive

Early forms of honey collecting entailed the destruction of the entire colony when the honey was harvested. The wild hive was crudely broken into, using smoke to suppress the bees, the honeycombs were torn out and smashed up — along with the eggs, larvae and honey they contained. The liquid honey from the destroyed brood nest was strained through a sieve or basket. This was destructive and unhygienic, but for hunter-gatherer societies this did not matter, since the honey was generally consumed immediately and there were always more wild colonies to exploit. But in settled societies the destruction of the bee colony meant the loss of a valuable resource; this drawback made beekeeping both inefficient and something of a "stop and start" activity. There could be no continuity of production and no possibility of selective breeding, since each bee colony was destroyed at harvest time, along with its precious queen.

Rural beekeeping in the 16th century

During the medieval period abbeys and monasteries were centers of beekeeping, since beeswax was highly prized for candles and fermented honey was used to make alcoholic mead in areas of Europe where vines would not grow. The 18th and 19th centuries saw successive stages of a revolution in beekeeping, which allowed the bees themselves to be preserved when taking the harvest.

Intermediate stages in the transition from the old beekeeping to the new were recorded for example by Thomas Wildman in 1768/1770, who described advances over the destructive old skep-based beekeeping so that the bees no longer had to be killed to harvest the honey. Wildman for example fixed a parallel array of wooden bars across the top of a straw hive or skep (with a separate straw top to be fixed on later) "so that there are in all seven bars of deal" [in a 10-inch-diameter (250 mm) hive] "to which the bees fix their combs." He also described using such hives in a multi-storey configuration, foreshadowing the modern use of supers: he described adding (at a proper time) successive straw hives below, and eventually removing the ones above when free of brood and filled with honey, so that the bees could be separately preserved at the harvest for a following season. Wildman also described a further development, using hives with "sliding frames" for the bees to build their comb, foreshadowing more modern uses of movable-comb hives. Wildman's book acknowledged the advances in knowledge of bees previously made by Swammerdam, Maraldi, and de Réaumur—he included a lengthy translation of Réaumur's account of the natural history of bees—and he also described the initiatives of others in designing hives for the preservation of bee-life when taking the harvest, citing in particular reports from Brittany dating from the 1750s, due to Comte de la Bourdonnaye. However, the forerunners of the modern hives with movable frames that are mainly used today are considered the traditional basket top bar (movable comb) hives of Greece, known as "Greek beehives". The oldest testimony on their use dates back to 1669 although it is probable that their use is more than 3000 years old.

Lorenzo Langstroth (1810–1895)

The 19th century saw this revolution in beekeeping practice completed through the perfection of the movable comb hive by the American Lorenzo Lorraine Langstroth. Langstroth was the first person to make practical use of Huber's earlier discovery that there was a specific spatial measurement between the wax combs, later called *the bee space*, which bees do not block with wax, but keep as a free passage. Having determined this bee space (between 5 and 8 mm, or 1/4 to 3/8"), Langstroth then designed a series of wooden frames within a rectangular hive box, carefully maintaining the correct space between successive frames, and found that the bees would build parallel honeycombs in the box without bonding them to each other or to the hive walls. This enables the beekeeper to slide any frame out of the hive for inspection, without harming the bees or the comb, protecting the eggs, larvae and pupae contained within the cells. It also meant that combs containing honey could be gently removed and the honey extracted without destroying the comb. The emptied honey combs could then be returned to the bees intact for refilling. Langstroth's book, *The Hive and Honey-bee*, published in 1853, described his rediscovery of the bee space and the development of his patent movable comb hive.

The invention and development of the movable-comb-hive fostered the growth of commercial honey production on a large scale in both Europe and the USA.

Evolution of Hive Designs

Langstroth's design for movable comb hives was seized upon by apiarists and inventors on both sides of the Atlantic and a wide range of moveable comb hives were designed and perfected in England, France, Germany and the United States. Classic designs evolved in each country: Dadant hives and Langstroth hives are still dominant in the USA; in France the De-Layens trough-hive became popular and in the UK a British National hive became standard as late as the 1930s although in Scotland the smaller Smith hive is still popular. In some Scandinavian countries and in Russia the traditional trough hive persisted until late in the 20th century and is still kept in some areas. However, the Langstroth and Dadant designs remain ubiquitous in the USA and also in many parts of Europe, though Sweden, Denmark, Germany, France and Italy all have their own national hive designs. Regional variations of hive evolved to reflect the climate, floral productivity and the reproductive characteristics of the various subspecies of native honey bee in each bio-region.

Bees at the hive entrance

The differences in hive dimensions are insignificant in comparison to the common factors in all these hives: they are all square or rectangular; they all use movable wooden frames; they all consist of a floor, brood-box, honey super, crown-board and roof. Hives have traditionally been constructed of cedar, pine, or cypress wood, but in recent years hives made from injection molded dense polystyrene have become increasingly important.

Honey-laden honeycomb in a wooden frame

Hives also use queen excluders between the brood-box and honey supers to keep the queen from laying eggs in cells next to those containing honey intended for consumption. Also, with the advent in the 20th century of mite pests, hive floors are often replaced for part of (or the whole) year with a wire mesh and removable tray.

Pioneers of Practical and Commercial Beekeeping

The 19th century produced an explosion of innovators and inventors who perfected the design and production of beehives, systems of management and husbandry, stock improvement by selective breeding, honey extraction and marketing. Preeminent among these innovators were:

Petro Prokopovych, used frames with channels in the side of the woodwork, these were packed side by side in boxes that were stacked one on top of the other. The bees travelling from frame to frame and box to box via the channels. The channels were similar to the cut outs in the sides of modern wooden sections (1814).

Jan Dzierżon, was the father of modern apiology and apiculture. All modern beehives are descendants of his design.

L. L. Langstroth, revered as the "father of American apiculture", no other individual has influenced modern beekeeping practice more than Lorenzo Lorraine Langstroth. His classic book *The Hive and Honey-bee* was published in 1853.

Moses Quinby, often termed 'the father of commercial beekeeping in the United States', author of *Mysteries of Bee-Keeping Explained.*

Amos Root, author of the *A B C of Bee Culture*, which has been continuously revised and remains in print. Root pioneered the manufacture of hives and the distribution of bee-packages in the United States.

A. J. Cook, author of *The Bee-Keepers' Guide; or Manual of the Apiary*, 1876.

Dr. C.C. Miller was one of the first entrepreneurs to actually make a living from apiculture. By 1878 he made beekeeping his sole business activity. His book, *Fifty Years Among the Bees*, remains a classic and his influence on bee management persists to this day.

Honey spinner

Major Francesco De Hruschka was an Italian military officer who made one crucial invention that catalyzed the commercial honey industry. In 1865 he invented a simple machine for extracting honey from the comb by means of centrifugal force. His original idea was simply to support combs in a metal framework and then spin them around within a container to collect honey as it was thrown out by centrifugal force. This meant that honeycombs could be returned to a hive undamaged but empty, saving the bees a vast amount of work, time, and materials. This single invention greatly improved the efficiency of honey harvesting and catalysed the modern honey industry.

Walter T. Kelley was an American pioneer of modern beekeeping in the early and mid-20th century. He greatly improved upon beekeeping equipment and clothing and went on to manufacture these items as well as other equipment. His company sold via catalog worldwide and his book, *How to Keep Bees & Sell Honey*, an introductory book of apiculture and marketing, allowed for a boom in beekeeping following World War II.

In the U.K. practical beekeeping was led in the early 20th century by a few men, pre-eminently Brother Adam and his Buckfast bee and R.O.B. Manley, author of many titles, including *Honey Production in the British Isles* and inventor of the Manley frame, still universally popular in the U.K. Other notable British pioneers include William Herrod-Hempsall and Gale.

Dr. Ahmed Zaky Abushady (1892–1955), was an Egyptian poet, medical doctor, bacteriologist and bee scientist who was active in England and in Egypt in the early part of the twentieth century. In 1919, Abushady patented a removable, standardized aluminum honeycomb. In 1919 he also founded The Apis Club in Benson, Oxfordshire, and its periodical Bee World, which was to be edited by Annie D. Betts and later by Dr. Eva Crane. The Apis Club was transitioned to the International Bee Research Association (IBRA). Its archives are held in the National Library of Wales. In Egypt in the 1930s, Abushady established The Bee Kingdom League and its organ, The Bee Kingdom.

In India, R. N. Mattoo was the pioneer worker in starting beekeeping with Indian honeybee, (*Apis cerana indica*) in early 1930s. Beekeeping with European honeybee, (*Apis mellifera*) was started by Dr. A. S. Atwal and his team members, O. P. Sharma and N. P. Goyal in Punjab in early 1960s.It remained confined to Punjab and Himachal Pradesh up to late 1970s. Later on in 1982, Dr. R. C. Sihag, working at Haryana Agricultural University, Hisar (Haryana), introduced and established this honeybee in Haryana and standardized its management practices for semi-arid-subtropical climates.On the basis of these practices, beekeeping with this honeybee could be extended to the rest of the country. Now beekeeping with *Apis mellifera* predominates in India.

Traditional Beekeeping

Wooden hives in Stripeikiai honeymaking museum, Lithuania

Beekeeping at Kawah Ijen Mountain, Indonesia

Fixed Comb Hives

A fixed comb hive is a hive in which the combs cannot be removed or manipulated for management or harvesting without permanently damaging the comb. Almost any hollow structure can be used for this purpose, such as a log gum, skep, wooden box, or a clay pot or tube. Fixed comb hives are no longer in common use in industrialized countries, and are illegal in places that require movable combs to inspect for problems such as varroa and American foulbrood. In many developing countries fixed comb hives are widely used and, because they can be made from any locally available material, are very inexpensive. Beekeeping using fixed comb hives is an essential part of the livelihoods of many communities in poor countries. The charity Bees for Development recognizes that local skills to manage bees in fixed comb hives are widespread in Africa, Asia, and South America. Internal size of fixed comb hives range from 32.7 liters (2000 cubic inches) typical of the clay tube hives used in Egypt to 282 liters (17209 cubic inches) for the Perone hive. Straw skeps, bee gums, and unframed box hives are unlawful in most US states, as the comb and brood cannot be inspected for diseases. However, skeps are still used for collecting swarms by hobbyists in the UK, before moving them into standard hives. Quinby used box hives to produce so much honey that he saturated the New York market in the 1860's. His writings contain excellent advice for management of bees in fixed comb hives.

Modern Beekeeping

Top-bar Hives

Top bar hives have been widely adopted in Africa where they are used to keep tropical honeybee ecotypes. Their advantages include being light weight, adaptable, easy to harvest honey, and less stressful for the bees. Disadvantages include combs that are fragile and cannot usually be extracted and returned to the bees to be refilled and that they cannot easily be expanded for additional honey storage.

A growing number of amateur beekeepers are adopting various top-bar hives similar to the type commonly found in Africa. Top bar hives were originally used as a traditional beekeeping method in Greece and Vietnam with a history dating back over 2000 years. These hives have no frames and the honey-filled comb is not returned after extraction. Because of this, the production of honey is likely to be somewhat less than that of a frame and super based hive such as Langstroth or Dadant. Top bar hives are mostly kept by people who are more interested in having bees in their garden than in honey production per se. Some of the most well known top-bar hive designs are the Kenyan Top Bar Hive with sloping sides, the Tanzanian Top Bar Hive with straight sides, and Vertical Top Bar Hives, such as the Warre or "People's Hive" designed by Abbe Warre in the mid-1900s.

The initial costs and equipment requirements are typically much less than other hive designs. Scrap wood or #2 or #3 pine can often be used to build a nice hive. Top-bar hives also offer some advantages to interacting with the bees and the amount of weight that must be lifted is greatly reduced. Top-bar hives are being widely used in developing countries in Africa and Asia as a result of the Bees for Development program. Since 2011, a growing number of beekeepers in the U.S. are using various top-bar hives.

Horizontal Frame Hives

The De-Layens hive, Jackson Horizontal Hive, and various chest type hives are widely used in Spain, France, Ukraine, Belarus, Africa, and parts of Russia. They are a step up from fixed comb and top bar hives because they have movable frames that can be extracted. Their limitation is primarily that volume is fixed and not easily expanded. Honey has to be removed one frame at a time, extracted or crushed, and the empty frames returned to be refilled. Various horizontal hives have been adapted and widely used for commercial migratory beekeeping. The Jackson Horizontal Hive is particularly well adapted for tropical agriculture. The De-Layens hive is popular in parts of Spain.

Vertical Stackable Frame Hives

In the United States, the Langstroth hive is commonly used. The Langstroth was the first successful top-opened hive with movable frames. Many other hive designs are based on the principle of bee space first described by Langstroth. The Langstroth hive is a descendant of Jan Dzierzon's Polish hive designs. In the United Kingdom, the most common type of hive is the British National, which can hold Hoffman, British Standard or Manley frames. It is not unusual to see some other sorts of hive (Smith, Commercial, WBC, Langstroth, and Rose). Dadant and Modified Dadant hives are widely used in France and Italy where their large size is an advantage. Square Dadant hives - often called 12 frame Dadant or Brother Adam hives - are used in large parts of Germany and other parts of Europe by commercial beekeepers. The Rose hive is a modern design that attempts to address many of the flaws and limitations of other movable frame hives. The only significant weakness of the Rose design is that it requires 2 or 3 boxes as a brood nest which infers a large number of frames to be worked when managing the bees. The major advantage shared by these designs is that additional brood and honey storage space can be added via boxes of frames added to the hive. This also simplifies honey collection since an entire box of honey can be removed instead of removing one frame at a time.

Protective Clothing

Most beekeepers also wear some protective clothing. Novice beekeepers usually wear gloves and a hooded suit or hat and veil. Experienced beekeepers sometimes elect not to use gloves because they inhibit delicate manipulations. The face and neck are the most important areas to protect, so most beekeepers wear at least a veil. Defensive bees are attracted to the breath, and a sting on the face can lead to much more pain and swelling than a sting elsewhere, while a sting on a bare hand can usually be quickly removed by fingernail scrape to reduce the amount of venom injected.

Beekeepers often wear protective clothing to protect themselves from stings

The protective clothing is generally light colored (but not colorful) and of a smooth material. This provides the maximum differentiation from the colony's natural predators (such as bears and skunks) which tend to be dark-colored and furry.

'Stings' retained in clothing fabric continue to pump out an alarm pheromone that attracts aggressive action and further stinging attacks. Washing suits regularly, and rinsing gloved hands in vinegar minimizes attraction.

Smoker

Smoke is the beekeeper's third line of defense. Most beekeepers use a "smoker"—a device designed to generate smoke from the incomplete combustion of various fuels. Smoke calms bees; it initiates a feeding response in anticipation of possible hive abandonment due to fire. Smoke also masks alarm pheromones released by guard bees or when bees are squashed in an inspection. The ensuing confusion creates an opportunity for the beekeeper to open the hive and work without triggering a defensive reaction. In addition, when a bee consumes honey the bee's abdomen distends, supposedly making it difficult to make the necessary flexes to sting, though this has not been tested scientifically.

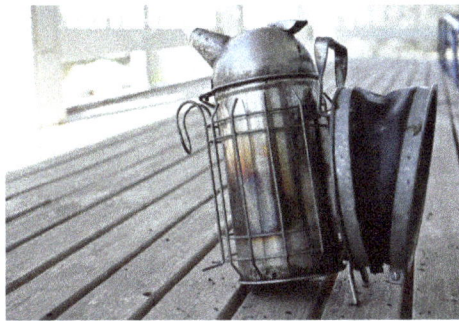

Bee smoker with heat shield and hook

Smoke is of questionable use with a swarm, because swarms do not have honey stores to feed on in response. Usually smoke is not needed, since swarms tend to be less defensive, as they have no stores or brood to defend, and a fresh swarm has fed well from the hive.

Many types of fuel can be used in a smoker as long as it is natural and not contaminated with harmful substances. These fuels include hessian, twine, burlap, pine needles, corrugated cardboard, and mostly rotten or punky wood. Indian beekeepers, especially in Kerala, often use coconut fibers as they are readily available, safe, and of negligible expense. Some beekeeping supply sources also sell commercial fuels like pulped paper and compressed cotton, or even aerosol cans of smoke. Other beekeepers use sumac as fuel because it ejects lots of smoke and doesn't have an odor.

Some beekeepers are using "liquid smoke" as a safer, more convenient alternative. It is a water-based solution that is sprayed onto the bees from a plastic spray bottle.

Torpor may also be induced by the introduction of chilled air into the hive – while chilled carbon dioxide may have harmful long-term effects.

Effects of Stings and of Protective Measures

Some beekeepers believe that the more stings a beekeeper receives, the less irritation each causes, and they consider it important for safety of the beekeeper to be stung a few times a season. Beekeepers have high levels of antibodies (mainly IgG) reacting to the major antigen of bee venom, phospholipase A2 (PLA). Antibodies correlate with the frequency of bee stings.

The entry of venom into the body from bee-stings may also be hindered and reduced by protective clothing that allows the wearer to remove stings and venom sacs with a simple tug on the clothing. Although the stinger is barbed, a worker bee is less likely to become lodged into clothing than human skin.

If a beekeeper is stung by a bee, there are many protective measures that should be taken in order to make sure the affected area does not become too irritated. The first cautionary step that should be taken following a bee sting is removing the stinger without squeezing the attached venom glands. A quick scrape with a fingernail is effective and intuitive. This step is effective in making sure that the venom injected does not spread, so the side effects of the sting will go away sooner. Washing the affected area with soap and water is also a good way to stop the spread of venom. The last step that needs to be taken is to apply ice or a cold compress to the stung area.

Natural Beekeeping

The natural beekeeping movement believes that modern beekeeping and agricultural practices, such as crop spraying, hive movement, frequent hive inspections, artificial insemination of queens, routine medication, and sugar water feeding, weaken bee hives.

Practitioners of 'natural beekeeping' tend to use variations of the top-bar hive, which is a simple design that retains the concept of movable comb without the use of frames or foundation. The horizontal top-bar hive, as championed by Marty Hardison, Michael Bush, Philip Chandler, Dennis Murrell and others, can be seen as a modernization of hollow log hives, with the addition of wooden bars of specific width from which bees hang their combs. Its widespread adoption in recent years can be attributed to the publication in 2007 of *The Barefoot Beekeeper* by Philip Chandler, which challenged many aspects of modern beekeeping and offered the horizontal top-bar hive as a viable alternative to the ubiquitous Langstroth-style movable-frame hive.

The most popular vertical top-bar hive is probably the Warré hive, based on a design by the French priest Abbé Émile Warré (1867–1951) and popularized by Dr. David Heaf in his English translation of Warré's book *L'Apiculture pour Tous* as *Beekeeping For All*.

Urban or Backyard Beekeeping

Related to natural beekeeping, urban beekeeping is an attempt to revert to a less industrialized way of obtaining honey by utilizing small-scale colonies that pollinate urban gardens. Urban apiculture has undergone a renaissance in the first decade of the 21st century, and urban beekeeping is seen by many as a growing trend.

Honey bee in Toronto

Some have found that "city bees" are actually healthier than "rural bees" because there are fewer pesticides and greater biodiversity. Urban bees may fail to find forage, however, and homeowners can use their landscapes to help feed local bee populations by planting flowers that provide nectar and pollen. An environment of year-round, uninterrupted bloom creates an ideal environment for colony reproduction.

Bee colonies

Castes

A colony of bees consists of three castes of bee:

- a queen bee, which is normally the only breeding female in the colony;

- a large number of female worker bees, typically 30,000–50,000 in number;

- a number of male drones, ranging from thousands in a strong hive in spring to very few during dearth or cold season.

Queen bee (center)

The queen is the only sexually mature female in the hive and all of the female worker bees and male drones are her offspring. The queen may live for up to three years or more and may be capable of laying half a million eggs or more in her lifetime. At the peak of the breeding season, late spring to summer, a good queen may be capable of laying 3,000 eggs in one day, more than her own body weight. This would be exceptional however; a prolific queen might peak at 2,000 eggs a day, but a more average queen might lay just 1,500 eggs per day. The queen is raised from a normal worker egg, but is fed a larger amount of royal jelly than a normal worker bee, resulting in a radically different growth and metamorphosis. The queen influences the colony by the production and dissemination of a variety of pheromones or "queen substances". One of these chemicals suppresses the development of ovaries in all the female worker bees in the hive and prevents them from laying eggs.

Mating of Queens

The queen emerges from her cell after 15 days of development and she remains in the hive for 3–7 days before venturing out on a mating flight. Mating flight is otherwise known as 'nuptial flight'. Her first orientation flight may only last a few seconds, just enough to mark the position of the hive. Subsequent mating flights may last from 5 minutes to 30 minutes, and she may mate with a number of male drones on each flight. Over several matings, possibly a dozen or more, the queen receives and stores enough sperm from a succession of drones to fertilize hundreds of thousands of eggs. If she does not manage to leave the hive to mate—possibly due to bad weather or being trapped in part of the hive—she remains infertile and become a *drone layer*, incapable of producing female worker bees. Worker bees sometimes kill a non-performing queen and produce another. Without a properly performing queen, the hive is doomed.

Mating takes place at some distance from the hive and often several hundred feet in the air; it is thought that this separates the strongest drones from the weaker ones, ensuring that only the fastest and strongest drones get to pass on their genes.

Worker Bees

Almost all the bees in a hive are female worker bees. At the height of summer when activity in the hive is frantic and work goes on non-stop, the life of a worker bee may be as short as 6 weeks; in late autumn, when no brood is being raised and no nectar is being harvested, a young bee may live for 16 weeks, right through the winter.

Female worker bee

Over the course of their lives, worker bees' duties are dictated by age. For the first few weeks of their lifespan, they perform basic chores within the hive: cleaning empty brood cells, removing debris and other housekeeping tasks, making wax for building or repairing comb, and feeding larvae. Later, they may ventilate the hive or guard the entrance. Older workers leave the hive daily, weather permitting, to forage for nectar, pollen, water, and propolis.

Period	Work activity
Days 1-3	Cleaning cells and incubation
Day 3-6	Feeding older larvae
Day 6-10	Feeding younger larvae
Day 8-16	Receiving nectar and pollen from field bees
Day 12-18	Beeswax making and cell building
Day 14 onwards	Entrance guards; nectar, pollen, water and propolis foraging; robbing other hives

Drones

Drones are the largest bees in the hive (except for the queen), at almost twice the size of a worker bee. They do not work, do not forage for pollen or nectar, are unable to sting, and have no other known function than to mate with new queens and fertilize them on their mating flights. A bee colony generally starts to raise drones a few weeks before building queen cells so they can supersede a failing queen or prepare for swarming. When queen-raising for the season is over, bees in colder climates drive drones out of the hive to die, biting and tearing their legs and wings.

Larger drones compared to smaller workers

Differing Stages of Development

Stage of development	Queen	Worker	Drone
Egg	3 days	3 days	3 days
Larva	8 days	10 days	13 days :Successive moults occur within this period 8 to 13 day period
Cell Capped	day 8	day 8	day 10
Pupa	4 days	8 days	8 days
Total	15 days	21 days	24 days

Structure of a Bee Colony

A domesticated bee colony is normally housed in a rectangular hive body, within which eight to ten parallel frames house the vertical plates of honeycomb that contain the eggs, larvae, pupae and food for the colony. If one were to cut a vertical cross-section through the hive from side to side, the brood nest would appear as a roughly ovoid ball spanning 5-8 frames of comb. The two outside combs at each side of the hive tend to be exclusively used for long-term storage of honey and pollen.

Within the central brood nest, a single frame of comb typically has a central disk of eggs, larvae and sealed brood cells that may extend almost to the edges of the frame. Immediately above the brood patch an arch of pollen-filled cells extends from side to side, and above that again a broader arch of honey-filled cells extends to the frame tops. The pollen is protein-rich food for developing larvae, while honey is also food but largely energy rich rather than protein rich. The nurse bees that care for the developing brood secrete a special food called 'royal jelly' after feeding themselves on honey and pollen. The amount of royal jelly fed to a larva determines whether it develops into a worker bee or a queen.

Apart from the honey stored within the central brood frames, the bees store surplus honey in combs above the brood nest. In modern hives the beekeeper places separate boxes, called 'supers', above the brood box, in which a series of shallower combs is provided for storage of honey. This enables the beekeeper to remove some of the supers in the late summer, and to extract the surplus honey harvest, without damaging the colony of bees and its brood nest below. If all the honey is 'stolen', including the amount of honey needed to survive winter, the beekeeper must replace these stores by feeding the bees sugar or corn syrup in autumn.

Annual Cycle of a Bee Colony

The development of a bee colony follows an annual cycle of growth that begins in spring with a rapid expansion of the brood nest, as soon as pollen is available for feeding larvae. Some production of brood may begin as early as January, even in a cold winter, but breeding accelerates towards a peak in May (in the northern hemisphere), producing an abundance of harvesting bees synchronized to the main nectar flow in that region. Each race of bees times this build-up slightly differently, depending on how the flora of its original region blooms. Some regions of Europe have two nectar flows: one in late spring and another in late August. Other regions have only a single nectar flow. The skill of the beekeeper lies in predicting when the nectar flow will occur in his area and in trying to ensure that his colonies achieve a maximum population of harvesters at exactly the right time.

The key factor in this is the prevention or skillful management of the swarming impulse. If a colony swarms unexpectedly and the beekeeper does not manage to capture the resulting swarm, he is likely to harvest significantly less honey from that hive, since he has lost half his worker bees at a single stroke. If, however, he can use the swarming impulse to breed a new queen but keep all the bees in the colony together, he maximizes his chances of a good harvest. It takes many years of learning and experience to be able to manage all these aspects successfully, though owing to variable circumstances many beginners often achieve a good honey harvest.

Formation of New Colonies

Colony Reproduction: Swarming and Supersedure

All colonies are totally dependent on their queen, who is the only egg-layer. However, even the best queens live only a few years and one or two years longevity is the norm. She can choose whether or not to fertilize an egg as she lays it; if she does so, it develops into a female worker bee; if she lays an unfertilized egg it becomes a male drone. She decides which type of egg to lay depending on the size of the open brood cell she encounters on the comb. In a small worker cell, she lays a fertilized egg; if she finds a larger drone cell, she lays an unfertilized drone egg.

A swarm about to land

All the time that the queen is fertile and laying eggs she produces a variety of pheromones, which control the behavior of the bees in the hive. These are commonly called *queen substance*, but there are various pheromones with different functions. As the queen ages, she begins to run out of stored sperm, and her pheromones begin to fail. Inevitably, the queen begins to falter, and the bees decide to replace her by creating a new queen from one of her worker eggs. They may do this because she has been damaged (lost a leg or an antenna), because she has run out of sperm and cannot lay fertilized eggs (has become a 'drone laying queen'), or because her pheromones have dwindled to where they cannot control all the bees in the hive.

At this juncture, the bees produce one or more queen cells by modifying existing worker cells that contain a normal female egg. However, the bees pursue two distinct behaviors:

1. Supersedure: queen replacement within one hive without swarming

2. Swarm cell production: the division of the hive into two colonies by swarming

Different sub-species of *Apis mellifera* exhibit differing swarming characteristics that reflect their evolution in different ecotopes of the European continent. In general the more northerly black races are said to swarm less and supersede more, whereas the more southerly yellow and grey varieties are said to swarm more frequently. The truth is complicated because of the prevalence of cross-breeding and hybridization of the sub species and opinions differ.

Supersedure is highly valued as a behavioral trait by beekeepers because a hive that supersedes its old queen does not swarm and so no stock is lost; it merely creates a new queen and allows the old one to fade away, or alternatively she is killed when the new queen emerges. When superseding a queen, the bees produce just one or two queen cells, characteristically in the center of the face of a broodcomb.

In swarming, by contrast, a great many queen cells are created—typically a dozen or more—and these are located around the edges of a broodcomb, most often at the sides and the bottom.

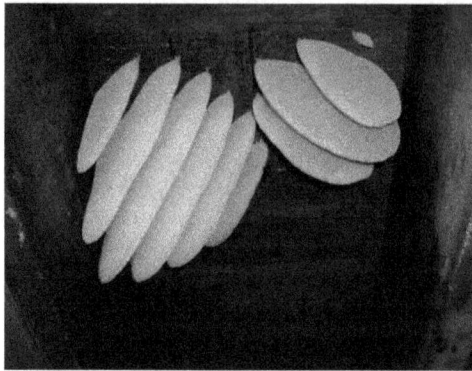

New wax combs between basement joists

Once either process has begun, the old queen normally leaves the hive with the hatching of the first queen cells. She leaves accompanied by a large number of bees, predominantly young bees (wax-secretors), who form the basis of the new hive. Scouts are sent out from the swarm to find suitable hollow trees or rock crevices. As soon as one is found, the entire swarm moves in. Within a matter of hours, they build new wax brood combs, using honey stores that the young bees have filled themselves with before leaving the old hive. Only young bees can secrete wax from special abdominal segments, and this is why swarms tend to contain more young bees. Often a number of virgin queens accompany the first swarm (the 'prime swarm'), and the old queen is replaced as soon as a daughter queen mates and begins laying. Otherwise, she is quickly superseded in the new home.

Factors that Trigger Swarming

It is generally accepted that a colony of bees does not swarm until they have completed all of their brood combs, i.e., filled all available space with eggs, larvae, and brood. This generally occurs in late spring at a time when the other areas of the hive are rapidly filling with honey stores. One key trigger of the swarming instinct is when the queen has no more room to lay eggs and the hive population is becoming very congested. Under these conditions, a prime swarm may issue with the queen, resulting in a halving of the population within the hive, leaving the old colony with a large number of hatching bees. The queen who leaves finds herself in a new hive with no eggs and no larvae but lots of energetic young bees who create a new set of brood combs from scratch in a very short time.

Another important factor in swarming is the age of the queen. Those under a year in age are unlikely to swarm unless they are extremely crowded, while older queens have swarming predisposition.

Beekeepers monitor their colonies carefully in spring and watch for the appearance of queen cells, which are a dramatic signal that the colony is determined to swarm.

When a colony has decided to swarm, queen cells are produced in numbers varying to a dozen or more. When the first of these queen cells is sealed after eight days of larval feeding, a virgin queen pupates and is due to emerge seven days later. Before leaving, the worker bees fill their stomachs with honey in preparation for the creation of new honeycombs in a new home. This cargo of honey also makes swarming bees less inclined to sting. A newly issued swarm is noticeably gentle for up to 24 hours and is often capable of being handled by a beekeeper without gloves or veil.

A swarm attached to a branch

This swarm looks for shelter. A beekeeper may capture it and introduce it into a new hive, helping meet this need. Otherwise, it returns to a feral state, in which case it finds shelter in a hollow tree, excavation, abandoned chimney, or even behind shutters.

Back at the original hive, the first virgin queen to emerge from her cell immediately seeks to kill all her rival queens still waiting to emerge. Usually, however, the bees deliberately prevent her from doing this, in which case, she too leads a second swarm from the hive. Successive swarms are called 'after-swarms' or 'casts' and can be very small, often with just a thousand or so bees—as opposed to a prime swarm, which may contain as many as ten to twenty-thousand bees.

A small after-swarm has less chance of survival and may threaten the original hive's survival if the number of individuals left is unsustainable. When a hive swarms despite the beekeeper's preventative efforts, a good management practice is to give the reduced hive a couple frames of open brood with eggs. This helps replenish the hive more quickly and gives a second opportunity to raise a queen if there is a mating failure.

Each race or sub-species of honey bee has its own swarming characteristics. Italian bees are very prolific and inclined to swarm; Northern European black bees have a strong tendency to supersede their old queen without swarming. These differences are the result of differing evolutionary pressures in the regions where each sub-species evolved.

Artificial Swarming

When a colony accidentally loses its queen, it is said to be "queenless". The workers realize that the queen is absent after as little as an hour, as her pheromones fade in the hive. The colony cannot survive without a fertile queen laying eggs to renew the population, so the workers select cells containing eggs aged less than three days and enlarge these cells dramatically to form "emergency queen cells". These appear similar to large peanut-like structures about an inch long that hang from the center or side of the brood combs. The developing larva in a queen cell is fed differently from an ordinary worker-bee; in addition to the normal honey and pollen, she receives a great deal of royal jelly, a special food secreted by young 'nurse bees' from the hypopharyngeal gland. This special food dramatically alters the growth and development of the larva so that, after metamorphosis and pupation, it emerges from the cell as a queen bee. The queen is the only bee in a colony which has fully developed ovaries, and she secretes a pheromone which suppresses the normal development of ovaries in all her workers.

Beekeepers use the ability of the bees to produce new queens to increase their colonies in a procedure called *splitting a colony*. To do this, they remove several brood combs from a healthy hive, taking care to leave the old queen behind. These combs must contain eggs or larvae less than three days old and be covered by young *nurse bees*, which care for the brood and keep it warm. These brood combs and attendant nurse bees are then placed into a small 'nucleus hive' with other combs containing honey and pollen. As soon as the nurse bees find themselves in this new hive and realize they have no queen, they set about constructing emergency queen cells using the eggs or larvae they have in the combs with them.

Diseases

The common agents of disease that affect adult honey bees include fungi, bacteria, protozoa, viruses, parasites, and poisons. The gross symptoms displayed by affected adult bees are very similar, whatever the cause, making it difficult for the apiarist to ascertain the causes of problems without microscopic identification of microorganisms or chemical analysis of poisons. Since 2006 colony losses from Colony Collapse Disorder have been increasing across the world although the causes of the syndrome are, as yet, unknown. In the US, commercial beekeepers have been increasing the number of hives to deal with higher rates attrition.

World Apiculture

Country	Production (1000 metric tons)	Consumption (1000 metric tons)	Number of beekeepers	Number of bee hives
World honey production and consumption in 2005				
Europe and Russia				
Ukraine	71.46	52		
Russia	52.13	54		
Spain	37.00	40		
Germany (*2008)	21.23	89	90,000*	1,000,000*
Hungary	19.71	4		
Romania	19.20	10		
Greece	16.27	16		
France	15.45	30		
Bulgaria	11.22	2		
Serbia	3 to 5	6.3	30,000	430,000
Denmark (*1996)	2.5	5	*4,000	*150,000
North America				
United States (*2006, **2002)	70.306*	158.75*	12,029** (210,000 bee keepers)	2,400,000*
Canada	45 (2006); 28 (2007)	29	13,000	500,000
Latin America				
Argentina	93.42 (Average 84)	3		
Mexico	50.63	31		
Brazil	33.75	2		
Uruguay	11.87	1		
Oceania				
Australia	18.46	16	12,000	520,000
New Zealand	9.69	8	2602	313,399
Asia				

China	299.33 (average 245)	238		7,200,000
Turkey	82.34 (average 70)	66		4,500,000
Iran				3,500,000
India	52.23	45		9,800,000
South Korea	23.82	27		
Vietnam	13.59	0		
Turkmenistan	10.46	10		
Africa				
Ethiopia	41.23	40		4,400,000
Tanzania	28.68	28		
Angola	23.77	23		
Kenya	22.00	21		
Egypt (*1997)	16*		200,000*	2,000,000*
Central African Republic	14.23	14		
Morocco (*1997)	4.5*		27,000*	400,000*
South Africa (*2008)	~2.5*	~1.5*	~1,790*	~92,000*
Source: Food and Agriculture Organization of the United Nations				

Sources:

- Denmark: beekeeping.com (1996)

- Arab countries: beekeeping.com (1997)

- USA: University of Arkansas National Agricultural Law Center, Agricultural Marketing Resource Center

- Serbia

References

- Cherry, Peter; Morris, T. R. (2008). Domestic Duck Production: Science and Practice. CABI. pp. 1–7. ISBN 978-1-84593-441-5.

- Smith, Andrew F. (2006). The Turkey: An American Story. University of Illinois Press. pp. 4–5, 17. ISBN 978-0-252-03163-2.

- Pond, Wilson, G.; Bell, Alan, W. (eds.) (2010). Turkeys: Behavior, Management and Well-Being. Marcell Dekker. pp. 847–849. ISBN 0-8247-5496-4.

- Crane, Eva The World History of Beekeeping and Honey Hunting, Routledge 1999, ISBN 0-415-92467-7, ISBN 978-0-415-92467-2.

- Dalley, S. (2002). Mari and Karana: Two Old Babylonian Cities (2 ed.). Gorgias Press LLC. p. 203. ISBN 978-1-931956-02-4.

- "Oldest known archaeological example of beekeeping discovered in Israel". Thaindian.com. 2008-09-01. Retrieved 2016-03-12.

- "A 17th Century Testimony On The Use Of Ceramic Top-bar Hives. 2012 | Haralampos Harissis and Georgios Mavrofridis". Academia.edu. 1970-01-01. Re-trieved 2016-03-12.

- "Economic aspects of beekeeping production in Croatia" (PDF). Veterinarski Arhiv. 79: 397–408. 2009. Retrieved 2016-03-12.

- Ingraham, Christopher (2015-07-23). "Call off the bee-pocalypse: U.S. honeybee colonies hit a 20-year high". The Washington Post. ISSN 0190-8286. Retrieved 2015-12-01.

- "Farm Commodity Programs: Honey" (PDF). nationalaglawcenter.org. National Honey Board. 2002. Retrieved 27 March 2014.

- Adler, Jerry; Lawler, Andrew (June 1, 2012). "How the Chicken Conquered the World". Smithsonian Magazine. Retrieved April 14, 2014.

- Buckland, R.; Guy, G. "Origins and Breeds of Domestic Geese". FAO Agriculture Department. Retrieved February 17, 2014.

- Viegas, Jennifer (February 1, 2010). "Native Americans First Tamed Turkeys 2,000 Years Ago". Retrieved February 19, 2014.

- Bolla, Gerry (April 1, 2007). "Primefacts: Squab raising" (PDF). NSW Department of Primary Industries. ISSN 1832-6668. Retrieved March 9, 2014.

- The Translational Genomics Research Institute (April 15, 2011). "US meat and poultry is widely contaminated with drug-resistant Staph bacteria, study finds". Science Daily. Retrieved February 27, 2014.

- "Information on Avian Influenza". Seasonal Influenza (Flu). Centers for Disease Control and Prevention. Retrieved March 3, 2014.

- "The evolution, and revolution, of flu vaccines". FDA: Consumer updates. U.S. Food and Drug Administration. January 18, 2013. Retrieved March 6, 2014.

- François Huber (1814). Nouvelles observations sur les abeilles,. Chez J. J. Paschoud, ... et a Geneve. Retrieved 27 March 2014.

- "USDA International Livestock & Poultry: World Duck, Goose and Guinea Fowl Meat Situation". The Poultry Site. December 19, 2013. Retrieved March 9, 2014.

- "Global Poultry Trends: World Egg Production Sets a Record Despite Slower Growth". The Poultry Site. January 16, 2013. Retrieved February 24, 2014.

Permissions

We would like to thank the editorial team for lending their expertise to make the book truly unique. They have played a crucial role in the development of this book. Without their invaluable contributions this book wouldn't have been possible. They have made vital efforts to compile up to date information on the varied aspects of this subject to make this book a valuable addition to the collection of many professionals and students.

This book was conceptualized with the vision of imparting up-to-date and integrated information in this field. To ensure the same, a matchless editorial board was set up. Every individual on the board went through rigorous rounds of assessment to prove their worth. After which they invested a large part of their time researching and compiling the most relevant data for our readers.

The editorial board has been involved in producing this book since its inception. They have spent rigorous hours researching and exploring the diverse topics which have resulted in the successful publishing of this book. They have passed on their knowledge of decades through this book. To expedite this challenging task, the publisher supported the team at every step. A small team of assistant editors was also appointed to further simplify the editing procedure and attain best results for the readers.

Apart from the editorial board, the designing team has also invested a significant amount of their time in understanding the subject and creating the most relevant covers. They scrutinized every image to scout for the most suitable representation of the subject and create an appropriate cover for the book.

The publishing team has been an ardent support to the editorial, designing and production team. Their endless efforts to recruit the best for this project, has resulted in the accomplishment of this book. They are a veteran in the field of academics and their pool of knowledge is as vast as their experience in printing. Their expertise and guidance has proved useful at every step. Their uncompromising quality standards have made this book an exceptional effort. Their encouragement from time to time has been an inspiration for everyone.

The publisher and the editorial board hope that this book will prove to be a valuable piece of knowledge for students, practitioners and scholars across the globe.

Index

www.ingramcontent.com/pod-product-compliance
Lightning Source LLC
Chambersburg PA
CBHW061933190326
41458CB00009B/2726